Biosystems & Biorobotics

Volume 23

Aims & Scope

Biosystems & Biorobotics publishes the latest research developments in three main areas: 1) understanding biological systems from a bioengineering point of view, i.e. the study of biosystems by exploiting engineering methods and tools to unveil their functioning principles and unrivalled performance; 2) design and development of biologically inspired machines and systems to be used for different purposes and in a variety of application contexts. The series welcomes contributions on novel design approaches, methods and tools as well as case studies on specific bioinspired systems; 3) design and developments of nano-, micro-, macrodevices and systems for biomedical applications, i.e. technologies that can improve modern healthcare and welfare by enabling novel solutions for prevention, diagnosis, surgery, prosthetics, rehabilitation and independent living.

On one side, the series focuses on recent methods and technologies which allow multiscale, multi-physics, high-resolution analysis and modeling of biological systems. A special emphasis on this side is given to the use of mechatronic and robotic systems as a tool for basic research in biology. On the other side, the series authoritatively reports on current theoretical and experimental challenges and developments related to the "biomechatronic" design of novel biorobotic machines. A special emphasis on this side is given to human-machine interaction and interfacing, and also to the ethical and social implications of this emerging research area, as key challenges for the acceptability and sustainability of biorobotics technology.

The main target of the series are engineers interested in biology and medicine, and specifically bioengineers and biroboticists. Volume published in the series comprise monographs, edited volumes, lecture notes, as well as selected conference proceedings and PhD theses. The series also publishes books purposely devoted to support education in bioengineering, biomedical engineering, biomechatronics and biorobotics at graduate and post-graduate levels.

About the Cover

The cover of the book series Biosystems & Biorobotics features a robotic hand prosthesis. This looks like a natural hand and is ready to be implanted on a human amputee to help them recover their physical capabilities. This picture was chosen to represent a variety of concepts and disciplines: from the understanding of biological systems to biomechatronics, bioinspiration and biomimetics; and from the concept of human-robot and human-machine interaction to the use of robots and, more generally, of engineering techniques for biological research and in healthcare. The picture also points to the social impact of bioengineering research and to its potential for improving human health and the quality of life of all individuals, including those with special needs. The picture was taken during the LIFEHAND experimental trials run at Università Campus Bio-Medico of Rome (Italy) in 2008. The LIFEHAND project tested the ability of an amputee patient to control the Cyberhand, a robotic prosthesis developed at Scuola Superiore Sant'Anna in Pisa (Italy), using the tf-LIFE electrodes developed at the Fraunhofer Institute for Biomedical Engineering (IBMT, Germany), which were implanted in the patient's arm. The implanted tf-LIFE electrodes were shown to enable bidirectional communication (from brain to hand and vice versa) between the brain and the Cyberhand. As a result, the patient was able to control complex movements of the prosthesis, while receiving sensory feedback in the form of direct neurostimulation. For more information please visit http://www.biorobotics.it or contact the Series Editor.

More information about this series at http://www.springer.com/series/10421

Athanasios Karafillidis · Robert Weidner
Editors

Developing Support Technologies

Integrating Multiple Perspectives to Create Assistance that People Really Want

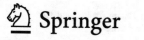
Springer

Editors
Athanasios Karafillidis
Laboratory of Manufacturing Technology
Helmut Schmidt University/University
  of the Federal Armed Forces Hamburg
Hamburg, Germany

Robert Weidner
Laboratory of Manufacturing Technology
Helmut Schmidt University/University
  of the Federal Armed Forces Hamburg
Hamburg, Germany

and

Chair of Production Technology
University of Innsbruck
Innsbruck, Austria

ISSN 2195-3562 ISSN 2195-3570 (electronic)
Biosystems & Biorobotics
ISBN 978-3-030-13198-2 ISBN 978-3-030-01836-8 (eBook)
https://doi.org/10.1007/978-3-030-01836-8

This Springer imprint is published by the registered company Springer Nature Switzerland AG
The registered company address is: Gewerbestrasse 11, 6330 Cham, Switzerland

Acknowledgements

This book is one of the many outcomes of a project that started in the end of 2014 at Helmut Schmidt University in Hamburg. The general objective of the project has been to gather expertise and knowledge for building technical support systems that people really want. This included a particular interest in physical support and the desire to overcome the predominant idea that automation and robotics unavoidably lead to a substitution of human labor.

The proposal to fund interdisciplinary competence in the field of human–machine interaction to face demographic change by the German Ministry of Education and Research (BMBF) provided an occasion to think different about technology development and its future challenges. In effect, the project smartASSIST was established with the generous support of the BMBF (grant no. 16SV7114). The project executing organization VDI/VDE Innovation + Technik GmbH took care of the necessary formal frame and helped to build up a network of collegiate research groups that emerged out of this research grant. The scientific advisors of our project, Klaus Henning and Philine Warnke, provided many helpful comments and encouraged the whole team to flesh out our conceptual and technological ideas. We are particularly indebted to Jens Wulfsberg, who hosts this project in his Laboratory of Manufacturing Technologies (Laboratorium Fertigungstechnik—LaFT), for providing the necessary research environment and for his steady trust in what we do.

The idea to publish this volume harks back to the second transdisciplinary conference on "Support Technologies that People Really Want" held at Helmut Schmidt University in Hamburg in December 2016. This book is a result of many discussions and exchange that happened during this conference (and beyond). We thank the university for supplying the necessary facilities and in particular the researchers, the staff, and all the other people who helped in one way or another to make it such a wonderful event. The process of compiling the book involves different dynamics, of course. Nothing of this work could have been done without all the authors who committed themselves to contribute to this volume. All of them are exceptional scholars, and we are grateful to have them in this book.

Andreas Argubi-Wollesen gave helpful comments on the introduction(s) and the last chapter. Also, we appreciate the courage of Springer to publish this unconventional book that spans multiple disciplines and perspectives.

Interdisciplinarity depends essentially on finding the right people to build up an exceptional culture of collaboration. This is a substantial experience we made in the course of this project. Therefore, we owe special thanks to Andreas Argubi-Wollesen, Jonas Klabunde, Christine Linnenberg, Bernward Otten, Tim Schubert, and Zhejun Yao of smartASSIST who bring this kind of collaboration to life each day.

Finally, we want to thank our families. All the research on support systems would not have been possible without these beautiful sociocultural and biophysical support systems that we both really want.

Athanasios Karafillidis
Robert Weidner

Contents

Part IV Values and Valuation

Prospects of a Digital Society

Contributors

Andrea Altepost Institute for Textile Engineering, RWTH Aachen University, Aachen, Germany

Kaspar Althoefer School of Engineering and Materials Science, Queen Mary University of London, London, UK

Andreas Argubi-Wollesen Laboratory of Manufacturing Technology, Helmut Schmidt University/University of the Federal Armed Forces Hamburg, Hamburg, Germany

Susanne Beck Faculty of Law, Leibniz University Hanover, Hanover, Germany

Peter Biniok Sociologist, Berlin, Germany

Andreas Bischof Junior Research Group "Miteinander", Media Informatics, University of Technology Chemnitz, Chemnitz, Germany

Laura L. Bischoff Institute of Human Movement Science, University of Hamburg, Hamburg, Germany

Gabriele Bleser Department of Computer Science, University of Kaiserslautern, Kaiserslautern, Germany

Sabrina Bringeland Institute of Human and Industrial Engineering, Karlsruhe Institute of Technology, Karlsruhe, Germany

Christina Bröhl Institute of Industrial Engineering and Ergonomics, RWTH Aachen University, Aachen, Germany

Barbara Deml Institute of Human and Industrial Engineering, Karlsruhe Institute of Technology, Karlsruhe, Germany

Annika Fohn Institute for Sociology, RWTH Aachen University, Aachen, Germany

Paul Glogowski Chair of Production Systems, Ruhr-University of Bochum, Bochum, Germany

Yves-Simon Gloy Institute for Textile Engineering, RWTH Aachen University, Aachen, Germany

Bruno Gransche Institute of Advanced Studies FoKoS, University of Siegen, Siegen, Germany

Christina M. Hein Micro Technology and Medical Device Technology, Technical University of Munich, Garching, Germany

Klaus Henning IMA/ZLW & IfU, P3 OSTO, RWTH Aachen University, Aachen, Germany

Jannis Hergesell Department of Sociology, DFG Graduate School "Innovation Society Today", Technical University of Berlin, Berlin, Germany

Daniel Houben Institute for Sociology, RWTH Aachen University, Aachen, Germany

Christoph Hubatschke Department of Philosophy, University of Vienna, Vienna, Austria

Alfred Hypki Chair of Production Systems, Ruhr-University of Bochum, Bochum, Germany

Athanasios Karafillidis Laboratory of Manufacturing Technology, Helmut Schmidt University/University of the Federal Armed Forces Hamburg, Hamburg, Germany

Bernd Kuhlenkötter Chair of Production Systems, Ruhr-University of Bochum, Bochum, Germany

Kai Lemmerz Chair of Production Systems, Ruhr-University of Bochum, Bochum, Germany

Kevin Liggieri Institute for Philosophy I, Ruhr University Bochum, Bochum, Germany

Janina Loh (née Sombetzki) Department of Philosophy, Philosophy of Technology and Media, University of Vienna, Vienna, Austria

Mario Löhrer Institute for Textile Engineering, RWTH Aachen University, Aachen, Germany

Tim C. Lueth Micro Technology and Medical Device Technology, Technical University of Munich, Garching, Germany

Paul Lukowicz German Research Center for Artificial Intelligence, Kaiserslautern, Germany

Arne Maibaum Department of Sociology, DFG Graduate School "Innovation Society Today", Technical University of Berlin, Berlin, Germany

Klaus Mattes Institute of Human Movement Science, University of Hamburg, Hamburg, Germany

Alexander Mertens Institute of Industrial Engineering and Ergonomics, RWTH Aachen University, Aachen, Germany

Jörg Miehling Engineering Design, Friedrich-Alexander-Universität Erlangen-Nürnberg, Erlangen, Germany

Christoph Müller Department for Architecture Theory and Philosophy of Technics, Vienna University of Technology, Vienna, Austria

Peter Müller Munich Center for Technology in Society/Digital Media Lab, Technical University of Munich, München, Germany

Kathrin Nuelle Institute of Mechatronic Systems, Leibniz University Hanover, Hanover, Germany

Tobias Ortmaier Institute of Mechatronic Systems, Leibniz University Hanover, Hanover, Germany

Bernward Otten Laboratory of Manufacturing Technology, Helmut Schmidt University/University of the Federal Armed Forces Hamburg, Hamburg, Germany

Kristin Paetzold Institute for Technical Product Development, University of the Federal Armed Forces Munich, Neubiberg, Germany

Jan-Hendrik Passoth Munich Center for Technology in Society/Digital Media Lab, Technical University of Munich, München, Germany

Peter Rasche Institute of Industrial Engineering and Ergonomics, RWTH Aachen University, Aachen, Germany

Arash Rezaey Institute for Textile Engineering, RWTH Aachen University, Aachen, Germany

Johannes Rönnfeldt Institute of Human Movement Science, University of Hamburg, Hamburg, Germany

Katharina Schäfer Institute of Industrial Engineering and Ergonomics, RWTH Aachen University, Aachen, Germany

Oliver Schürer Department for Architecture Theory and Philosophy of Technics, Vienna University of Technology, Vienna, Austria

Benjamin Stangl Department for Architecture Theory and Philosophy of Technics, Vienna University of Technology, Vienna, Austria

Agostino Stilli Department of Computer Science, University College London, London, UK

Bertram Taetz Department of Computer Science, University of Kaiserslautern, Kaiserslautern, Germany

Svenja Tappe Institute of Mechatronic Systems, Leibniz University Hanover, Hanover, Germany

Sabine Theis Institute of Industrial Engineering and Ergonomics, RWTH Aachen University, Aachen, Germany

Sandro Wartzack Engineering Design, Friedrich-Alexander-Universität Erlangen-Nürnberg, Erlangen, Germany

Karsten Weber Institute for Social Research and Technology Assessment (IST), Ostbayerische Technische Hochschule (OTH), Regensburg, Germany

Robert Weidner Laboratory of Manufacturing Technology, Helmut Schmidt University/University of the Federal Armed Forces Hamburg, Hamburg, Germany; Chair of Production Technology, University of Innsbruck, Innsbruck, Austria

Rebecca Wiczorek Department of Psychology and Ergonomics, Technical University of Berlin, Berlin, Germany

Matthias Wille Institute of Industrial Engineering and Ergonomics, RWTH Aachen University, Aachen, Germany

Alexander Wolf Engineering Design, Friedrich-Alexander-Universität Erlangen-Nürnberg, Erlangen, Germany

Bettina Wollesen Institute of Human Movement Science, University of Hamburg, Hamburg, Germany

Helge A. Wurdemann Department of Mechanical Engineering, University College London, London, UK

Zhejun Yao Laboratory of Manufacturing Technology, Helmut Schmidt University/University of the Federal Armed Forces Hamburg, Hamburg, Germany

Introduction

Developing. Support. Technologies.

Athanasios Karafillidis and Robert Weidner

The relationship of humans and technology has changed significantly during the last two decades. It has become closer and multiplex—that is, technology has moved closer to the human body and their interconnections have become entangled and diverse. In this vein, technology is envisioned as being able to support or assist human beings more profoundly than ever before. This change has occurred incrementally and is still in progress. It has led to a diversification of possible application areas, a shift in the landscape of innovation projects, and a different perception of technological possibilities in general.

The reasons for this change are manifold. Big societal trends like globalization, individualization, disruptive technological innovations, and demographic pressures are no doubt important for explaining the change in human–technology relations. All of them push political agendas and channel research funds. But when it comes to an understanding of these transformations, they only yield a universal interpretive frame. The recently enforced closeness and multiplexity of human–technology relations are much better understood when taking into account the expanding connectivity, the distribution and increase of computational power, and the plummeting costs of material components and production. Their combination has a high impact on further technological possibilities, but also on the perceptions, needs, and expectations related to technology.

However, this is still only a part of the story. The shift of the relevant relationships is not only a response to such technical or social pressures outside of innovation projects. It is also an inside job. The myriad micro-processes unfolding in the world's

A. Karafillidis (✉) · R. Weidner
Laboratory of Manufacturing Technology, Helmut Schmidt University/University of the Federal Armed Forces Hamburg, Hamburg, Germany
e-mail: karafillidis@hsu-hh.de

R. Weidner
Chair of Production Technology, University of Innsbruck, Innsbruck, Austria
e-mail: robert.weidner@hsu-hh.de

© Springer Nature Switzerland AG 2018
A. Karafillidis and R. Weidner (eds.), *Developing Support Technologies*, Biosystems & Biorobotics 23, https://doi.org/10.1007/978-3-030-01836-8_1

engineering laboratories, in thousands of projects, and in the corporations' research and development departments day by day bring forth different and new relations of humans and technologies. Political claims, technological possibilities, funding interests, scientific progress, legal regulations, cultural images, or market-driven demands no doubt influence them deeply. But societal expectations and demands with regard to technology are themselves profoundly shaped by the very form of organization, the underlying beliefs, and the approaches used in the relevant projects. Thus, the internal reason, as it were, for the observable shift is the advent of a different style of developing technology. This book will argue that the idea of *support* dwells at the center of this shift. It has spawned technological devices that are conceptualized and built with the explicit intention to serve the needs of individuals at work and in everyday life. The objective of this volume is to expound how this is done in detail and what needs to be considered to be able to proceed in a successful as well as responsible way.

Usually, this is considered as the domain of human–machine interaction (HMI). Yet this approach is not equipped to handle the practical, ethical, social, and also technical issues involved. The main lines of this classic program are confined to the interaction of separate units—albeit we are dealing with different forms of entanglement already that blur habitual boundaries and give rise to hybrid entities. Thus, the crucial difference made by a technology that is expected to support, assist, or help people remains obscure when simply seen through the lens of HMI. In contrast, conceiving any human–machine or human–computer interaction and collaboration as support relation gives way to a new approach that certainly draws on the rich tradition of HMI but advances beyond its limitations.

Up to now, discussions about support technologies—from autonomous robots to monitoring systems, wearables, and implants, both in forms of simple tools and complex gadgetry—are preoccupied with engineering issues. It is supposed that if the invented devices are intended to support people, they will do so automatically. This is misleading. No engineering and no design can stipulate the purpose of a device uniquely and unequivocally. A technical invention has to be understood as only one part of a wider support system that also comprises organizational, physical, ethical, legal, and cognitive components. Therefore, the future challenge in research, construction, implementation, and deployment of such systems is twofold. On the one hand, theories and concepts have to be developed accordingly in order to generate fresh, diverse, and surprising perspectives on the relevant problems. On the other hand, new methods and forms of collaboration and evaluation are required to integrate and implement these ideas—in other words, to provide the requisite contexts in which they might grow and get the chance to be cherished and become successful.

This book brings together scholars from heterogeneous disciplines and research fields like biomechanics, engineering, social science, psychology, law, and philosophy to meet this twofold challenge. It integrates both different conceptual perspectives and issues of interdisciplinary development in one volume and sometimes even within single chapters. Since it tries to account for the complex and intertwined technical, social, cognitive, and ethical contexts of technology development and design, this volume gives an idea of how responsible research and innovation is currently realized in developing support technologies. Strictly speaking, the whole endeavor is not about bringing technology to the people. It is about finding ways to design and evaluate technology in tune with the people so that it finds its way to the them in the course of the process—and vice versa.

***

"Developing Support Technologies" is not just some title for this book but rather signifies a research program in a nutshell. The following explanation of the ideas and associations of this title will unfold its main characteristics and substantiate the book's main purpose as well as some of its contents and contentions.

1 Developing—A New Field, a Form of Design/Construction, and Transdisciplinarity

"Developing Support Technologies" has a double meaning that must be considered. On the one hand, "developing" has an active meaning in the sense of bringing technologies forth by designing/constructing, building, and evaluating them. This refers to the setting that there are teams of developers, mostly within departments of profit or non-profit organizations, who work on the implementation of concrete technical solutions for certain specified applications and are preoccupied with managing technical uncertainties [Moh17, Ger15]. On the other hand, "developing" refers to a developing field, a new and therefore necessarily incomplete and sometimes fragile strand of technology and its accompanying societal and organizational uncertainties. This refers to support technologies as a developing area within society in general and engineering in particular [Ois10, Bin17].

The twofold understanding of the title unites the two scientific cultures of engineering on the one hand and the social sciences and humanities on the other. Depending on their scientific background, people read the title in one of these two different meanings. The observable division of work between them does also transpire roughly along these lines. Social science and humanities are more focused on the developing field or certain parts of it and ascertain the accompanying structures and their societal embeddings, while engineering pays more attention to finding feasible and reproducible technical principles that lead to viable, reliable, and tangible material results. Although this kind of division of labor and interest with regard to technology exists, these two groups of disciplines have been getting closer recently since it became obvious that support technologies need to function technically as well as socioculturally. A device that does not work in a technical sense will not be accepted, but a device that triggers fear or requires specialized knowledge or clothing will not "work" either—or will only be accepted under certain conditions or by particular groups. To be sure, this holds for any technology, but the advent of support technologies (also called assistance technologies prematurely) has altered the relevant perspectives. The sociocultural embedding of technology has now become explicit. It is not only accounted for in hindsight but before and during development. The social sciences and humanities are now considered as an important part of the development process more frequently—many remaining obstacles notwithstanding [Vis15].

To develop technology proper has been and still is, to be sure, the work of engineers. They conceive, design, construct, test, and improve composite devices, machines, and systems until they meet the defined requirements. But when it comes to support technologies, this classic form of development is expanded. First, the focus on people with their impression on what counts and their expression of demands alters the technical search for suited materials, proper joints, or adequate programming and favors quickly adaptable devices and simple, intelligible controls. However, the requisite requirements can not be defined once and for all. They are refined and redefined during development on many levels [Suc07, p. 278]. Second, the sequence of development from conception to implementation is not fixed or linear anymore but supplanted by parallel and circular processes. Classic waterfall models of project management have not disappeared, but they fulfill a different function: They provide a rough orientation and legitimize the approach vis-à-vis third parties, like investors and funding agencies. Yet progress in the actual everyday work of research and development is not achieved by sticking to some plan. Mixing up the diverse structures and managing their dissonance allows for an organizational responsiveness [Sta09] that is part and parcel of developing support systems. Third, an augmentation of classic development occurs simply because many more diverse people are involved than before—with various disciplinary backgrounds but also without any academic interest: especially potential users, corporate groups, businesses, social media publics, and further stakeholders.

In short, developing support technologies involve/involves participation, interdisciplinarity, and new organizational forms [Bro15]. The true concept for this threefold augmentation of classical development processes is *transdisciplinarity*: New forms of collaboration have to be found between various scientific disciplines and also between them and potential users, interested citizens, and project partners. To sum up, speaking of *developing* support technologies entails a double perspective and transdisciplinary approaches to collaboration.

2 Support—Different Forms, Structural Properties, and Antithesis to Substitution

The term "support" might sound a little odd in times when most technologies in this vein are labeled as "assistive." Most of the time these two are used synonymously, but their difference matters. In our conception, support is the generic term. Assistance and help are particular subcategories of support that require distinctive situational structures [Kar17]. Assistance occurs when a task is divided into subtasks and the situationally participating entities, e.g., human individuals and technical devices, are assigned different subtasks in a complementary fashion. Think of the assistant of a CEO or the assistant devices built into modern cars as examples of such a complementary form of support. A tool, in contrast, does not "assist" its user. Even just saying it sounds awkward. Yet there is no doubt that a tool provides support. Also, a mother does not "assist" her children but supports or even helps them wherever possible.

Whether a gadget assists or helps is negotiable and proves to be crucial for its design and acceptance. An exoskeleton enabling a movement that otherwise would be impossible does not assist the individual but rather helps, because it is granted the control over the activity to constitute the movement after all—when help is understood as a form of support that passes the control of the activity in question to the helping entity. When a different exoskeleton is designed to decrease musculoskeletal stress, it supports an activity that could also be performed without it. In this case, neither exactly "help" nor "assistance" do apply for a proper characterization of the unfolding process. Still, it is providing support.

Two general structural properties of support situations (comprising assistance and help) are observer dependence and asymmetric relations. If technology development is attentive to human needs and societal acceptance, it must not ignore that the provision of support lies in the eye of the beholder. This concerns questions about who or what is supporting whom but also whether somebody is assigned support or rather asks for it. Interests, interpretations, and contextual conditions of the observing agencies (i.e., individuals, groups, organizations or other social systems, maybe robots) might lead either to a fierce rejection or to a passionate use of the support system. The other point in case is asymmetry. The development of support technologies must be aware that any support introduces some asymmetry between the supporting and the supported unit, which cannot simply be programmed and settled purposefully in advance [Suc07, pp. 268–269]. Which form of asymmetry prevails in the end can only be identified in frequent field tests and painstaking observation of real use cases in the wild. Often, the time factor of development projects thwarts such a thorough investigation. Yet, first steps into this direction can be clearly recognized. The importance of iteration and external feedback for constructing both responsibly according to human demands and successfully with respect to acceptance has already been realized indeed—albeit there is still way to go for a wider acknowledgment.

In addition to the distinction of different forms of support and the structural properties of support situations, there is a third significant aspect of the "support" component in the book title. Support is, as it were, the antithesis to substitution. Until recently, the sometimes hidden but mainly overt curriculum and objective of technology development has been automation, that is, the substitution of human workforce by machines. The reasons were economic in most cases, like increased productivity, effectivity, and efficiency but also better product quality, less mistakes, improved ergonomic conditions, and not least, to be sure, honorable ambitions to spare humans doing dangerous, risky, and strenuous work.

Many positive effects of automation could be enumerated, and the downsides are also well known [For15, Car14]. It remains a moot point, whether the positive and negative aspects of automation balance each other or not. However, there is an intriguing issue that is more constructive in nature than any debates about loss or gain of jobs. It concerns less the paradoxes, glitches, or unintended effects of automation but rather its pragmatic and practical limits. Certainly, there are a lot of engineers and entrepreneurs who expect that someday any task and activity can be automated. Maybe they will be proven right some day, but this is not the case in point at all. The crucial question is, how we should invest our time and resources to find adequate solutions for the limits and problems we *currently* face.

Engineering research in automation remains important and will no doubt continue. But it should not happen in expense of finding solutions that integrate the skills and awareness of humans with suitable technology. It is important to note that this "integration" exceeds the ideas of interaction and collaboration prevailing in automation engineering. Investing in support technologies receives rightly more attention because when it comes to developing new technology, the substitutionalist paradigm has reached certain limits. Many tasks and activities will not be amenable to automation for a long time to come. For example, anybody who has experienced existing robots for elderly care does immediately recognize this. By implication, human beings will remain pivotal for value creation in plants as well as other formal organizations. Physical skills and human awareness will become even more important in future value chains. Whatever the hopes projected into some indeterminate future: with respect to complex assembly tasks, evaluation, sensorimotor skills, judgment, discretion, the recognition of opportunities, diversity, quick adaptability, heedful perception of weak signals, or situational awareness human beings will remain indispensable.

In short, developing support technologies is/are contingent on a closer inspection of the structural conditions of support and their subtle nuances. Support is the proper answer to the practical limits of automation and to the substitution of human work. To sum up, developing *support* technologies entails to realize that the creation of value and values as well as the evaluation of situations and opportunities cannot abstain from cognition, that is, both awareness and sensorimotor skills.

3   Technologies—Innovation, Customization, and Modularity

Support systems transcend mere technological devices that are devised to assist or support human activities. Yet, the focus on developing support *technologies* is crucial. Having a techno-material structure available or at least imagining some materially tangible device allows to summon all interested participants effectively [Sta89]. It provides a powerful pretense to think about the sociotechnical design of support systems. Other existing forms of support in society—like neighborhood help, emotional and financial support, or assistive functions in hierarchies—lack this summoning material "thing." Presumably, this corresponds to the lack of studies that examine the internal structure of support situations more closely. Studies of "social support" [Hou88] have been preoccupied with structural conditions and individual effects of providing support and less with the inner functioning and patterning of support situations proper. Considering support from the perspective of *technology* urges new perspectives on structure and process of support in general.

Technologies transcend isolated devices. They involve networks of people, skills, material things, certain knowledge, and particular stances to the world [Mac99]. The technical devices in these networks seem like reliable islands of functioning causality that are intimately integrated in a scaffolding of unreliable components. Thus, the concept of "techno-logy"—and not simply: "technics." The Greek word *logos* implies a specific form of reason that accompanies the technics (though not necessarily philosophical reason as a universal). There is always a peculiar logic that pervades technical gadgets and their causal functioning. Next to this "grammar" of technology, there is also a pragmatic knowledge. The term techno-logy indicates that knowledge about construction principles, interaction, and handling is an integral part of its functioning. Finally, technologies are surrounded by a particular wording before, during, and after their development: for example, by justifications and poli-cies, the engineering and design parlance, or the typical marketing vocabulary. In this vein, techno-logy incorporates the syntax, pragmatics, and semantics of a causally constructed material structure that is expected to produce certain determined effects repeatedly and reliably [Ram07, p. 45].

Exactly this societal embedding of technologies also distinguishes mere *inventions* that constantly pop up in laboratories, garages, and institutions on the one hand and durable, accepted, and disseminated structures called *innovations* on the other [Ram10]. Any path to innovation needs to be paved through the muddy grounds of society. Previously, this has been done unconsciously and in passing. In developing support technologies, this aspect is brought to mind explicitly and allows to account for it from the outset. This is not a guarantee for innovation and success but nonetheless gives new design options and some leverage in a process that has been considered stochastic so far.

Technologies are both about *products* and *production*. To develop support tech-nologies means to develop products that people and organizations really want. At the same time, it means to develop technologies that are deployed in the production of products and become part of value chains and production processes—not only in the conventional sense of industrial production but also in the unconventional, generic sense of production, which includes the production of services and private DIY production. The switch to the idea of support tightens the intimate connection of these two aspects.

One of the most salient effects of the support paradigm is the approaching of the human body by technology [Vis03]. This makes any technical product also acces-sible, deployable, and potentially beneficial for production processes, which are transformed in consequence. From there, new forms of technical products and even innovations can arise. The distant machine hall in which products are produced far removed from everyday life is losing the importance it had since the industrial rev-olution. The German term "Industrie 4.0," cyber-physical systems, sociotechnical systems, or digitization of production are all expressions of this transformation. Citizen science, the democratization of production, or the character of the "prosumer" characterize the same issues from the opposite side. The same smart gadgets that are used in daily life are now integrated into organized work processes and production plants. An exoskeleton might help the residents of a nursing home to maintain some autonomy but can also be used by the personnel to manage their work load and reduce physical strain.

The proceeding (mass) *customization* of products can likewise be linked to the idea of support. The major response to customization demands is no doubt automation. Today, it is possible to specify the own preferences for a product online and to thereby trigger an automated process in some machine park to produce the desired product that is then automatically packaged and dispatched. Three issues are important in this scenario. First, support technologies need not necessarily operate in the proximity of human individuals. They can be distributed over time and space. Second, support does not necessarily preclude substitution. In the described case the customer is supported to design its own product (within certain limits) while the corporation substitutes human workforce. Third, there are moments in this automated process, for example, quality control, that are difficult to automate. Furthermore, there are also sectors where customization depends completely on human skill, for example, the construction and adaptation of prostheses, many forms of surgery, haircuts, all forms of nursery, or most products of construction industry, especially the completion of the interior. The people involved in such customization procedures are already supported by proper software or (smart as well as classical) tools but there is much more potential, in particular with respect to physical support.

A last facet of this unfoldment of the volume title is the necessary plural of "technologies" in connection to support. It should have become clear that support itself cannot be automated. It is the customized product per se because it involves and generates hybrid entities that merge certain activities, human bodies, technical devices, perceptions, norms, and social situations. That is, there will always exist *many* suitable support solutions for diverse activities and contexts—but also for seemingly identical activities and contexts. The latter points to another crucial engineering challenge in developing support technologies: to achieve *adaptability* and *modularity*. Since the hybrid combination of the human body, its perceptional capabilities, and technical equipment has to be considered as unique in every situation, the standardization prospects are disappointing. Support technologies are thus drivers of devising new forms of technical adaptability to bodies, perceptions, and situations with the objective to form *one integrated system*.

This adaptability can also be achieved by inventing modular solutions. Modularity, however, also refers to a more intriguing, though very challenging, aspect: the modularity and customization of support that is achieved by coupling different technical systems which in turn requires the construction, standardization, and design of compatible interfaces for different components of support systems. That is, for example, various interfaces for signal and information transfer, energy transmission, physical contact surfaces, or handling and control. Both variability within interfaces and between interfaces are highly relevant.

In short, support technologies are embedded in a whole apparatus of non-technical components and rely on them to function properly. Innovations come from finding a proper fit between all of these components. To sum up the consequence, developing support *technologies* pose/poses some challenges for engineering. They compel the profession to rethink the connection of products and production, to reconsider the relation of automation and customization, and to develop adaptable and modular systems as well as relevant interfaces that make them compatible.

***

The contributions in this book display the diversity of the people, disciplines, and topics in this field of research. Not all of the above aspects of the presented research program are discussed in this publication. However, the diversity is explicit, and its management is not an easy task. This includes the editing of the book. We selected distinguished scholars that are not only experts in their respective field but who additionally have some experience with participatory and interdisciplinary research projects regarding technology development for support. As editors we had a general concept for the book in mind and targeted the relevant researchers to send us articles treating a particular subfield, presenting subject-specific views on support systems, or reporting about deployment contexts from the perspective of their expertise.

The chapters that made it into the book are grouped into four major parts: "Demands and Expectations," "Constructing and Construing," "Forms and Contexts of Deployment," and "Values and Valuation." They are followed by two concluding chapters that discuss some prospective further developments of (support) technologies. The four parts represent main clusters of research activities in developing support technologies. They seem to form a sequence, but this is owed to the book format only. More likely, they set up a circular process. Starting with a demand analysis seems natural, but there are already valuations in place or earlier prototypes that lead to certain demands. Furthermore, all of these activities run concurrently in relevant projects and permanently influence each other. This implies that none of these discrete yet interfaced "stages" is ever completed as long as the project unfolds. Demand analysis is an ongoing concern in the development of support technologies as are valuation, construction, and deployment.

The main ideas framing each subsection and short introductions to the individual chapters are given in brief introductory notes at the beginning of the book parts. Due to the just mentioned circularity and simultaneity of the empirical processes, there are overlaps. Some of the articles could appear in more than one subsection. Yet there are good reasons to arrange them this way. One of the editorial decision premises in this respect has been to demonstrate the multiplicity of perspectives within each research cluster as represented by the subsections of this volume. We have decisively refrained from making special sections for, e.g., science, engineering, and humanities. There are no leading disciplines in any of the research areas for developing support technologies.

***

The papers collected here come from many different disciplines. The volume contains inputs from biomechanics, engineering, information science, philosophy, psychology, and the social sciences—to name only the most generic denominations and sparing the internal specializations. Each of the chapters cherishes its own terminology and quirkiness. All of them, however, can also be read by researchers who are not familiar with the subject-specific debates of the disciplines. The language they use is generally intelligible. Despite that, no article is able to deny its origin and background. Such a denial or disguise of disciplines would have been detrimental to the idea of this book. A seminal reference between disciplines is only possible when the *difference* between them is retained and accepted.

Certainly, to some extent this book displays the personal, regional, and institutional networks of the editors and their research group. But this Central European bias is not simply accidental. The *transdisciplinary* research community on technical support systems is actually prevalent in Central Europe. This may be due to research policy decisions or some other factors not yet explored. In the end, it may be just our ignorance. But there is no doubt that an international publication putting the common thread of these multiple perspectives on developing support technologies into focus is overdue. In this respect, this book is also an appeal asking for further international communication and collaboration in this developing research field—and also an appeal to prove us wrong that the form of transdisciplinary research on support technologies as expounded in this introduction is mainly happening in Europe (see, however, [Ois10] and [San14] with similar ambitions). Any further information and suggestions are welcome by the editors as well as all of the contributors.

This collection and arrangement of research papers is not the classic "how to" book. It contains reflections and descriptions of the different processes that accompany any development of support technologies. Its effect is a change in perspective and this generates, then again, we hope, a plethora of ideas how to approach and implement one's own projects. Therefore, it is a book of research in two respects. First, it allows to observe and thus research how support technologies are developed; and second, it gives some leverage to do research based on these suggestions and experiences from others.

References

[Bin17] Biniok, P., & Lettkemann, E. (Eds.). (2017). *Assistive Gesellschaft. Multidisziplinäre Erkundungen zur Sozialform "Assistenz"*. Wiesbaden: Springer VS.

[Bro15] Brown, R. R., Deletic, A., & Wong, T. H. F. (2015). How to catalyse collaboration. *Nature, 525*, 315–317.

[Car14] Carr, N. (2014). *The glass cage. How our computers are changing us*. New York: W. W. Norton.

[For15] Ford, M. (2015). *Rise of the robots. Technology and the threat of a jobless future*. New York: Basic Books.

[Ger15] Gerke, W. (2015). *Technische Assistenzsysteme. Vom Industrieroboter zum Roboterassistenten*. Berlin et al.: Walter de Gruyter.

[Hou88] House, J. S., Umberson, D., & Landis, K. R. (1988). Structures and processes of social support. *Annual Review of Sociology, 14*, 293–318.

[Kar17] Karafillidis, A. (2017) Synchronisierung, Kopplung und Kontrolle in Netzwerken. Zur sozialen Form von (technischer) Unterstützung und Assistenz. In P. Biniok & E. Lettkemann (Eds.), *Assistive Gesellschaft* (pp. 27–58). Wiesbaden: Springer VS.

[Mac99] MacKenzie, D. & Wajcman, J. (Eds.). (1999). The social shaping of technology (2nd ed.). Buckingham: Open UP.

[Moh17] Mohammed, S., Park, H. W., Park, C. H., Amirat, Y., & Argall, B. (2017). Special issue on assistive and rehabilitation robotics. *Autonomous Robots, 41*(3), 513–517.

[Ois10] Oishi, M. M. K., Mitchell, I. A., & Van der Loos, H. F. M. (Eds.). (2010). *Design and use of assistive technologies. Social, technical, ethical, and economic challenges*. New York et al.: Springer.

[Ram07] Rammert, W. (2007). *Technik – Handeln – Wissen. Zu einer pragmatischen Technik- und Sozialtheorie*. Wiesbaden: VS Verlag.

[Ram10] Rammert, W. (2010). Die Innovationen der Gesellschaft. In J. Howaldt & H. Jacobsen (Eds.), *Soziale Innovation. Auf dem Weg zu einem postindustriellen Innovationsparadigma* (pp. 21–51). Wiesbaden: VS Verlag.

[San14] Sankai, Y., Suzuki, K., & Hasegawa, Y. (Eds.). (2014). *Cybernics. Fusion of human, machine and information systems*. Springer Japan.

[Sta89] Star, S. L., & Griesemer, J. R. (1989). Institutional ecology, 'translations' and boundary objects: amateurs and professionals in Berkeley's Museum of Vertebrate Zoology, 1907–1939. *Social Studies of Science, 19*, 387–420.

[Sta09] Stark, D. (2009). *The sense of dissonance. Accounts of worth in economic life*. Princeton: Princeton UP.

[Suc07] Suchman, L. (2007). *Human-machine reconfigurations. Plans and situated actions* (2nd ed.). Cambridge: Cambridge UP.

[Vis03] Viseu, A. (2003). Simulation and augmentation: Issues of wearable computers. *Ethics and Information Technology, 5*, 17–26.

[Vis15] Viseu, A. (2015). Integration of social science into research in crucial. *Nature, 525*, 291.

Part I
Demands and Expectations

Technology is pervaded by narratives that justify the effort of their development. A permanent feature of such inevitable narratives is the presentation of technical solutions as responses to some demand or need. This feature is reflected, for example, in the "motivation" of engineers or in the well-known requirement specifications. Yet it makes a difference where the recounted demands come from. Most ideas still emerge out of what is technically feasible and then look for external demands to which they appear as an answer. To be sure, this does not mark a problem per se. Such a technology-driven practice has its own edge. But starting with an analysis of situated and domain-specific demands, no doubt makes a difference—in particular when acceptance is an issue.

Developing technical systems in response to demands is the first and crucial step to increase the probability of their acceptance. Demands of potential users and stakeholders can be surveyed by observing people and practices, body movements and task environments, routines and interactions. Various methods exist to get the requisite observations, e.g., diverse interview techniques, experimental setups in the laboratory, participant observation, field tests, or ethnographies. To yield an expedient input for engineering, the gathered data is used to reconstruct the multiple conditions of work and life, in which the contrived technology is to be integrated. Potentials, possibilities, and risks of a support technology entering the users' worlds can be induced from there.

Although demand analysis is rightly understood to mark the beginning of some project, it is also important to realize that it is an ongoing accomplishment. Demands and needs are not stable but shift in time. They change when a prototype comes into play, when technology is utilized or deployed in other contexts, or after it is evaluated and further optimized.

All in all, demands—including needs, acceptance, and usability issues—represent the *expectations* in the relevant field and explain its dynamics. The expectations of stakeholders are heterogeneous and do often differ from those of the prospective users, and both in turn differ from those of the involved journalists, managers, or politicians. Any analysis of demands (and thus acceptance) is contingent on the field of further expectations connected to the solutions and the project in general. Thus, demands and acceptance should not be treated separately. Their analysis goes hand in hand. Both are part of the web of expectations surrounding, informing, and driving the respective project.

This first subsection of the book summarizes different approaches from diverse disciplines to expound what is necessary for a suitable analysis of demands and expectations in developing support technologies. In detail, they are considered with regard to sociotechnical arrangements, human practices, meaning, bodily mechanics, societal trends, and the phenomenal worlds we perceive and inhabit.

Peter Biniok starts with a piece that sets the general stage. He argues that any development of support technology takes place in "sociotechnical assistance ensembles." He stresses the importance of fine-grained negotiation practices in such arrangements and describes relevant processes and structural features. Practitioners thus get an impression of the overall expectational dynamics in a project and gain an understanding of how seemingly "soft" factors constitute "hard" technology and become vital for the success of the technical solution and the endeavor as a whole.

Kristin Paetzold continues with the presentation of a technique that can be used to analyze needs of (older) people to find leverage points for technical support. The proposed practice-centered analysis of needs takes into account the local and material contexts of older people's immediate lifeworlds and then looks at the practical strategies they develop to deal with everyday problems. Any need for support can be understood with more precision when routines and practices are observed in local environments. The gained insights can be put to use for product development, as Paetzold argues in conclusion.

Kevin Liggieri opens up a historical perspective that helps to understand the role of meaning in relations of humans and technology. Searching for technical structures with a high probability of acceptance, the early discipline of psychotechnics had contrived the concept of "Sinnfälligkeit" which described how technical and material structures made sense to human users when they matched their mental and corporeal structures. Technical systems have to make sense, to fall into place, and to generate meaning. What is now called "intuitive control" does not only have historical predecessors, but the account of Liggieri also highlights that the acceptance of technology is a question of embodied meaning and sense-making.

With the contribution of *Andreas Argubi-Wollesen* and *Robert Weidner*, we move on from sociotechnical, material, and cultural demands to understanding and measuring demands of the human body. The authors present the distinctive challenges posed by the human body and its movements for the development of physical support technologies, i.e., exoskeletons, and give an overview of expedient tools. They contend that there is no single best method but only a suitable mixture of different methods that have to be selected and recombined for each research and development project anew.

Local and situated demand analysis for support technologies can be complemented by harnessing, as it were, the law of large numbers. *Alexander Mertens, Katharina Schäfer, Sabine Theis, Christina Bröhl, Peter Rasche,* and *Matthias Wille* shift the perspective to such large-scale measurements of societal dynamics and trends. The mass survey techniques presented by the authors help practitioners to identify general demands belonging to a certain social group (users, customers) and to understand how these technological demands and expectations change over time.

At the end of this first part of the book, *Bruno Gransche* inquires into how technical support/assistance is finding its way into our lifeworlds and how this reshapes our relations to technology. His arguments point to an aspect that is ignored in technology development most of the time, namely, that assistance is a thoroughly two-sided affair. This is not only a reminder that unintended consequences may emerge. Instead, it addresses the very practical issue that any analysis of expectations and demands should be extended to include the expectations of developers and stakeholders. Gransche exemplifies this by showing that the prevalent idea of bringing relief to the people by support technologies remains flawed if the burdens of assistance are ignored.

What is a factual relief for one group of people might be a burden for another; what is a relief at one point in time might turn out as a burden some time later; and what gives relief in the performance of one task might appear as a burden when performing a different task. This inevitable two-sidedness of expectations is not a "nice-to-have" insight. Rather, it has severe practical consequences for issues of acceptance and must be taken into account to develop sound, sustainable, and successful technical support systems that people really want.

Sociotechnical Assistance Ensembles. Negotiations of Needs and Acceptance of Support Technologies

Peter Biniok

Abstract Questions of needs and acceptance of support technologies are negotiated in sociotechnical assistance ensembles. Sociotechnical assistance ensembles are discussed as analytical tool, which overcomes both the dichotomy of social issues and technology as well as the distinction of technology development and use. In this way, it is possible to take all relevant heterogeneous actors (humans as well as technologies) into account and explore their relationships during technization procedures. The development of support technologies is at best a participatory, multidisciplinary, transformative, and ecological process.

1 Introduction: Assistance, Technology, and Society

Present society is (also) an assistive society [Bin17a][1]. *Assistance* is observable in various forms, such as care and cooperation among people (laboratory assistant), help by simple technologies (navigation device) or collaboration in complex configurations of social actors and technical entities (computer-assisted surgery). Especially the latter form of ever closer exchange between humans and support technologies has steadily been increasing in the past few years. This affects not only the use and distribution (*acceptance*), but also the design and development (*needs*) of support technologies. Technical sociology [Ram07] and science and technology studies [Hac07] deal with these processes of innovation, technization and utilization, and investigate the interdependencies between humans and technologies.

On the one hand, technologies are shaped by social, cultural, economic and other factors (social constructivism). For instance, depending on the local conditions in laboratories, technologies vary over place and time, even if the basic functions are the same. On the other hand, technologies effect society (technological determinism). They enable or prevent actions of people, e.g., doors grant access and/or stop thieves.

P. Biniok (✉)
Sociologist, Berlin, Germany
e-mail: peter.biniok@freenet.de

[1]This work is based on the more detailed proceeding [Bin16a].

© Springer Nature Switzerland AG 2018
A. Karafillidis and R. Weidner (eds.), *Developing Support Technologies*, Biosystems & Biorobotics 23, https://doi.org/10.1007/978-3-030-01836-8_2

These two perspectives are joined in the concept of *technopragmatism*. This approach starts from the assumption that humans and technologies are always in interaction and depend on each other. Since technology has agency, both humans and technologies shape society continuously in relationships of distributed action. Following this perspective, I discuss the concept of sociotechnical assistance ensembles as an analytical tool to investigate support situations and support technologies. The fields of ambient assisted living (AAL) and technological supported care illustrate in this work the discrepancies between claims of development *and* actual result as well as between the stated *and* observed research practices. One important aspect here is the divergence between assumptions about the needs of people and their actual demands. Another crucial question is how support technologies developed in laboratories and under science-oriented criteria may fit in the complex life worlds of self-determined individuals. The advantage of the suggested approach is that humans and technologies are understood as being interrelated and mutually dependent and have to adapt to each other. The analytical focus is on configurations of concrete actions and interactions in particular, but shifts according to research questions to groups, organizations and systems. In this way, the two separated spheres of technology development and use are perceived as one unit of analysis. Support technologies are not built in isolation from users and the context of usage, and they are not used independent from design departments and laboratories. Sociotechnical assistance ensembles focus on *support technologies as subject of negotiation* between heterogeneous actors from the first generation of ideas, via design and development to the use of technology.

2 Sociotechnical Assistance Ensembles and Negotiations of Technical Support

The concept of sociotechnical assistance ensembles is proposed for a comprehensive description, analysis and implementation of support technologies. Sociotechnical assistance ensembles are situational, goal-driven, and well-planned configurations of heterogeneous instances, i.e. people, machines, or materials [Bin17b]. *Assistance* is understood as a distributed practice: one or more provider of assistance conduct assistance services to one or more users. The provision of assistance enables the user(s) to perform specific actions and thus to continue a course of action. Assistance providers may be humans, technologies, organizations, institutions, or natural phenomena. The specificity of *sociotechnical* assistance ensembles is that technologies play an active role as a smart and/or intelligent action partner (e.g. algorithms and software agents).

Support technologies are an integral part of sociotechnical assistance ensembles [Wei15]: they are focal objects of the design and development processes as well as the main component of use practices. This means that various actors discuss and negotiate the appearance and functionality of support technologies, their specific parameters, requirements, and areas of application. Both, need and acceptance are dependent on situations and contexts. Decisions for technical support that are based solely on engineer visions and/or pure user centering do not refer to real world needs nor lead to acceptance of technologies in everyday action. Questions of needs and acceptance of technical support are not (only) answered on the user and application level, but much earlier and much later in sociotechnical assistance ensembles from the start of the idea of a support technology to its usage in real life on site.

3 Needs of Support Technologies: Exploration of Lifeworlds

Technology development is not only oriented toward efficiency and rationality, but at the same time always influenced by prevailing technological paradigms, sociotechnical regimes, discourses, and societal imagery. The social shaping of technology [Mac99] depends on the vision of the "internet of things" and slogans like "active aging", cost issues and user preferences, material properties, legal regulations of data protection, etc. Alliances, negotiations and power relations will have an impact on the final outcome of the technization processes [Lat98]. The analytical focus of sociotechnical assistance ensembles on the early stages of technology development includes support technologies, designers and engineers as well as actors from science policy, research paradigms, and societal images. In addition, future users and scholars from the social sciences are participating to inquire about the different needs of various life worlds. The development of support technologies is a *participatory and multidisciplinary* process.

3.1 Participation: Scripts and Frames

Users exist long before a technology is developed and used [Oud03]. Already at the beginning of development processes, teams asses and imagine future users and their motives to use a support technology. On the one hand, techno-scientific knowledge and implicit assumptions and expectations of designers and developers are incorporated in ideas and concepts. On the other hand, data derives from market research or from requirements ascertained from user surveys. Both perspectives—but especially the latter one—are taken into account to develop technologies that are needed.

During the conception, design and development stages, these ideas, assumptions, and functionalities are inscribed into technologies as 'scripts' [Akr92]. Scripts are implicit instructions on how to use a device. The greater the fit between a script (as anticipated user) and the real user, the more a technology meets a need. The difficulty is that expectations of technical support depend on the point of view of any relevant actor. Depending on the frames of reference, the development will result in different forms of technical support. Studies point to the existence of multiple requests that have to be harmonized. At the same time, the studies reveal considerable discrepancies in the reference frames of developers and users and/or between the assumptions about the benefits of technical assistance in the lifeworlds of the elderly and workplaces of caregivers [Kri17]. Other studies show that scripts of technologies hardly address the future user group [End16, End17]. This can be ascribed to an insufficient requirements analysis or an inadequate implementation of functionalities. For example, young computer scientists develop a sensor to recognize if older people fall. But only on the basis of their own experience and without a user survey the development will likely fail.

To match scripts, frames and needs, users have to be participants. *Sociotechnical assistance ensembles* reflect not only user deficiencies and compensatory functions of assistive technologies, they widen the scope to scripts, frames, and "hidden" perceptions from developers and investors. To investigate these scripts and the needs of primary and secondary users, more sciences, such as social sciences, become part of the development processes.

3.2 Multidisciplinary: Expertise and Competence

In the early stages of technology development, the reflection of scripting is essential, and the inclusion of social scientists is needed to deal with the specific needs and reference frames. Users have to participate in technization projects but user centration in the form of (standard) surveys gives insights only into "typical" problems that do not cover the complexity of society. What is needed is a more *subject-oriented* perspective. Technology development then follows the paradigm of participation of potential users in the development processes. However, the gaps between the different social worlds of design, development, science and civil society exist. Various restrictions, also appear in participatory design. Studies show that developers ignore critical evaluation results and/or only selectively consider research findings from other disciplines [Com15]. In addition, user tests to verify technologies may not be representative and test series in laboratories are hardly everyday case studies. Possible side effects and unintended effects of the use of technologies are not detected.

Precisely here, cooperation between designers, technicians, and social scientists is useful. Every discipline has its own methods and approaches to capture user experiences, to generate new ideas and/or to identify needs and wishes of users. But too often, the role of the social sciences as "accompanying research" is limited to the ex-post-legitimation of supporting technologies. My own participation in the AAL-projects showed that the different types of expertise in research processes need much more recognition. More attention should be paid to a *suitable* labor division and transfer management. Both, the focus on lifeworlds as a whole and continuous involvement of at least the primary users by means of interviews, workshops, group discussions, and observations is necessary [Bin16b]. The aim is the multidimensional participation of all relevant actors from science and civil society to go beyond 'consumerism' and allow 'empowerment' by technologies. *Sociotechnical assistance ensembles* refer to these challenges and provide an approach to knowledge-based divisions of labor and the establishment of multidisciplinary standards.

4 Acceptance of Support Technologies: Embed in Daily Routine

Acceptance comes not just from the working and handling of a technology. Acceptance is moreover related to individual preferences, social values, societal structures, and so on. Many studies, especially those related to risk and science communication, aim to capture the attitude of people towards innovation processes and products to reduce prejudices and to query potentials. This is observable in the field of supporting technologies too. But this perspective has to be complemented by analyzing technologies in everyday use. Laboratory settings do not have the necessary complexity of real-world conditions. Users assign technologies with specific meaning precisely in everyday life [Pel16]. This means that early tests in practice over a long period of time should be performed and evaluated on-site. Such tests provide knowledge about how further actors, such as service providers (education) or institutions (law), influence the utilization processes or are influenced by them. In addition, non-technical expertise is necessary to process and coordinate evaluation procedures. Sociotechnical assistance ensembles enable the investigation of this form of acceptance of support technologies. It follows that technology development is at best *transformative and ecological*.

4.1 Transformation: Trust and Practice

Technology in use remains a black box, because it is not necessary for users to have knowledge about its operation. When users are confident that devices work, they accept a technology. This acceptance is based for example on the brand of the manufacturer, the corporate culture, and/or on standardization [Wag94]. However, these factors only matter if technologies are available on the market. For many supporting technologies, this is not (yet) the case. On the actual acceptance of AAL-technologies or care robots or the like, little is known so far because both broad acceptance as well as longitudinal studies are missing [Kün15]. What is known are the factors influencing acceptance such as ease-of-use, costs, and technical affinity. Acceptance of supporting technology is achieved if use may lead to improvements of personal development or self-realization, whereas connections to negatively connoted age images or motives of deficit compensation lead to rejection of technologies [Pel16].

Crucial for the acceptance of (new) assistive technologies is the support of users in adoption and usage processes. Sociotechnical assistance ensembles expand the view from proper operation to areas of operation and point to the relevance of service phones, operating instructions, standards, but also to the consideration of ethical and legal aspects. Studies on accessibility of computer technologies show that (intro-ductory) courses and informally trying out of technologies promote their acceptance [Bin15]. The knowledge of a competent contact person give users the necessary security while engaging with technology. In other cases, acceptance of assistance has to be generated by demonstration [Tre17]. The demonstration of the abilities of a kitchen robot in a laboratory setting helps an audience to understand support technologies. The demonstration shows what the robot is able to learn and generates the assumption that the robot is a companion in everyday life. Assistance then is negotiated between developers, technology, and the audience. *Sociotechnical assis-tance ensembles* deal with the use and training of support technologies and with their allocation of meaning.

4.2   Ecology: Placing and Adjustment

During practice and training of support technologies changes in everyday life may occur. Research of acceptance attempts to reflect and estimate the impact of tech-nologies on the users' lives. Technologies restructure for example social relations and communication structures [Hör95]. Individuals follow scripts of devices and machines and set up their daily practice—at the same time they (re)interpret them according to their individual preferences. Moreover, technologies affect human actions directly by giving instructions and fulfilling tasks in cooperation with them [Ram02]. For a successful application of technologies, real-world conditions have to be taken into account, as in the form of an "eco-check": Are all necessary infrastruc-tures (social, technological) available? Is anyone else affected if people use this or that technology? What changes on site can be expected in case of long-term usage?

Practical tests of supporting technologies over a long period of time in every-day life show that their use leads to new options for action and interaction, e.g. the increase of social participation of senior citizens [Bin15]. New computer tech-nologies and new skills enable people to contact relatives living abroad via video telephony and deepen existing social contacts in the region. Other studies show that negotiations take place to fit the technologies into existing social patterns. For exam-ple, despite the adoption of computer technologies by a wife, the roles of technology competence in a household-relationship remain stable [Die17]. The wife "just" con-tinues to take over communication tasks with her tablet-PC (operation knowledge), whereas the husband keeps the tasks of computer oriented problem-solving with a PC (functionality knowledge). And technologies are changing working places too. Studies show that the introduction of new safety technologies not only change the behavior of people with dementia, but also affect the entire configuration of actors of care [Her17]. The new technologies lead to care as restrictive practice, which had been otherwise situationally negotiated. The perspective of *sociotechnical assistance ensembles* demonstrates that roles and meanings of support technologies are generated in practice, and acceptance depends on existing structures, power relations, and interests.

5   Technological Development and Use Are Inseparable

Needs and acceptance of support technologies are the result of subjective *negotiations between heterogeneous actors* instead of objective determinants and findings. What a need is, is determined by the actual wishes of users, depends on the skills of scientists and designers, and the resistance of technologies. In this process, it is important to exploit the lifeworlds and workplaces of users as well as to question the laboratory worlds of developers and engineers and their scientific paradigms. The determination of isolated needs runs the risk of overlooking the complex reference frames of support technologies. The relationship between needs and acceptance is ambivalent. The successful addressing of a need with a technology does not imply the acceptance of it. Acceptance of a technology does not depend solely on its functionality but on its successful integration in everyday life. This includes associated infrastructures (such as Internet access) and financing possibilities just as the emancipation of users, meaning readiness-to-hand of devices and having background knowledge about them.

Technology development happens in the nexus of action and structure [Bin13]: technization as a collective process takes place in relation to sociotechnical contexts and distributed actions of humans and technologies. Technology and social issues cannot be separated. Likewise, technology development and use can hardly be separated. They have to be supposed and realized as one process, because the realms of development and use are interconnected. Sociotechnical assistance ensembles are an analytical concept to direct the focus on humans *and* technologies, on development *and* use of technologies, and moreover on technical *and* social science. The development of support technologies is a participatory, multidisciplinary, transformative and ecological process. This implies that new types of research projects are necessary: long term, inter-institutional, holistic. Therewith lies the possibility to increase and enhance people's quality of life by means of support technologies.

References

[Akr92]   Akrich, M. (1992). The de-scription of technical objects. In W. E. Bijker & J. Law (Eds.), *Shaping technology/building society: Studies in sociotechnical change* (pp. 205–224). Cambridge: MIT Press.

[Bin13]   Biniok, P. (2013). Technische Entwicklung im Nexus von Handlung und Struktur. *EWE, 24,* 520–523.

[Bin15]   Biniok, P., & Menke, I. (2015). Societal participation of the elderly: Information and communication technologies as a "Social Junction". *Anthropology and Aging, 36,* 164–181.

[Bin16a]   Biniok, P. (2016). Soziotechnische Assistenzensembles. Aushandlungen von Bedarf und
           Akzeptanz technischer Unterstützungssysteme. In R. Weidner (Ed.), Zweite Transdiszi-
           plinäre Konferenz "Technische Unterstützungssysteme, die die Menschen wirklich wol-
           len" (pp. 269–283). Hamburg.
[Bin16b]   Biniok, P., Menke, I., & Selke, S. (2016). Social inclusion of elderly people in rural areas
           by social and technological mechanisms. In E. Domínguez-Rué & L. Nierling (Eds.),
           Ageing and technology (pp. 93–117). Transcript: Bielefeld.
[Bin17a]   Biniok, P., & Lettkemann, E. (Eds.). (2017). Assistive Gesellschaft. Multidisziplinäre
           Erkundungen zur Sozialform "Assistenz". Wiesbaden: Springer VS.
[Bin17b]   Biniok, P., & Lettkemann, E. (2017). In Gesellschaft – Assistenzformen, Assisten-
           zweisen und Assistenzensembles. In P. Biniok & E. Lettkemann (Eds.), Assistive
           Gesellschaft (pp. 1–23). Wiesbaden: Springer VS.
[Com15]   Compagna, D., & Kohlbacher, F. (2015). The limits of participatory technology devel-
           opment: The case of service robots in care facilities for older people. Technological
           Forecasting and Social Change, 93, 19–31.
[Die17]   Dietel, K. (2017). Generations- und geschlechtsspezifische Technikaneignung im tech-
           nikunterstützen Wohnen. In P. Biniok & E. Lettkemann (Eds.), Assistive Gesellschaft
           (pp. 225–249). Wiesbaden: Springer VS.
[End16]   Endter, C. (2016). Skripting age—The negotiation of age and aging in ambient assisted
           living. In E. Domínguez-Rué & L. Nierling (Eds.), Ageing and technology (pp. 121–140).
           Transcript: Bielefeld.
[End17]   Endter, C. (2017). Assistiert altern. Die Entwicklung eines Sturzsensors im Kontext von
           Ambient Assisted Living. In P. Biniok & E. Lettkemann (Eds.), Assistive Gesellschaft
           (pp. 167–181). Wiesbaden: Springer VS.
[Hac07]   Hackett, E. J., Amsterdamska, O., Lynch, M. E., & Wajcman, J. (Eds.). (2007). The
           handbook of science and technology studies. Cambridge: MIT Press.
[Her17]   Hergesell, J. (2017). Assistive Sicherheitstechniken in der Pflege von an Demenz
           erkrankten Menschen. In P. Biniok & E. Lettkemann (Eds.), Assistive Gesellschaft
           (pp. 203–223). Wiesbaden: Springer VS.
[Hör95]   Hörning, K. H. (1995). Technik und Kultur: ein verwickeltes Spiel der Praxis. In J. Half-
           mann, G. Bechmann, & W. Rammert (Eds.), Technik und Gesellschaft (pp. 131–151).
           Campus: Frankfurt a.M.
[Kri17]   Krings, B.-J., & Weinberger, N. (2017). Kann es technische Assistenten in der Pflege
           geben? Überlegungen zum Begriff der Assistenz in Pflegekontexten. In P. Biniok & E.
           Lettkemann (Eds.), Assistive Gesellschaft (pp. 183–201). Wiesbaden: Springer VS.
[Kün15]   Künemund, H. (2015). Chancen und Herausforderungen assistiver Technik. Nutzerbe-
           darfe und Technikakzeptanz im Alter. TATuP, 24, 28–35.
[Lat98]   Latour, B. (1998). Eine neue Politik der Dinge und für die Menschen. Innovationen in
           Technik. In E. Fricke (Ed.), Wissenschaft und Gesellschaft (pp. 147–181). Bonn: FES.
[Mac99]   MacKenzie, D. A., & Wajcman, J. (1999). The social shaping of technology. Bucking-
           ham: Open University Press.
[Oud03]   Oudshoorn, N., & Pinch, T. (2003). How users matter. The co-construction of users and
           technology. Cambridge: MIT Press.
[Pel16]   Pelizäus-Hoffmeister, H. (2016). Motives of the elderly for the use of technology in
           their daily lives. In E. Domínguez-Rué & L. Nierling (Eds.), Ageing and technology
           (pp. 27–46). Transcript: Bielefeld.
[Ram02]   Rammert, W., & Schulz-Schaeffer, I. (Eds.). (2002). Können Maschinen handeln? Sozi-
           ologische Beiträge zum Verhältnis von Mensch und Technik. Frankfurt a. M.: Campus.
[Ram07]   Rammert, W. (2007). Technik – Handeln – Wissen. Zu einer pragmatischen Technik- und
           Sozialtheorie. Wiesbaden: VS Verlag.

[Tre17]   Treusch, P. (2017). Humanoide Roboter als zukünftige assistive Akteure in der Küche?
          In P. Biniok & E. Lettkemann (Eds.), *Assistive Gesellschaft* (pp. 251–274). Wiesbaden:
          Springer VS.
[Wag94]   Wagner, G. (1994). Vertrauen in Technik. *Zeitschrift für Soziologie, 23,* 145–157.
[Wei15]   Weidner, R., Redlich, T., & Wulfsberg, J. P. (Eds.). (2015). *Technische Unter-
          stützungssysteme*. Wiesbaden: Springer VS.

Context-Integrating, Practice-Centered Analysis of Needs

Kristin Paetzold

Abstract The development of supporting technology often neglects real challenges for a self-determined lifestyle, especially in age. The objective of the contribution is to explain the KPB-methodology, which was developed in a project founded by the German Ministry of Education and Research (BMBF). This methodology allows capturing needs and problem situations of elderly people in their domestic environment. Implications for product development will be explained. With the project, we answered the question, which problems elderly people have to deal with for a self-determined life. We investigated the life situation based on socio-scientific methods and translated it into technical requirements.

1 Introduction

Nowadays, it is undisputed that technical systems can help human beings to overcome naturally given performance limits. Just think of the capability of flying. Irrespective of whether the performance limits are caused by biological factors, illness or age, technical systems in their various forms can help to strengthen own resources and expand the options for action. Activities in both the professional and private environment can be carried out in a time- and energy-optimized manner, which in turn creates space for personal self-realization. However, this requires that the technical systems are perceived and accepted by the user regarding to these possibilities.

K. Paetzold (✉)
Institute for Technical Product Development, University of the Federal Armed Forces Munich, Werner-Heisenberg-Weg 39, 85577 Neubiberg, Germany
e-mail: kristin.paetzold@unibw.de

© Springer Nature Switzerland AG 2018
A. Karafillidis and R. Weidner (eds.), *Developing Support Technologies*, Biosystems & Biorobotics 23, https://doi.org/10.1007/978-3-030-01836-8_3

2 Responsibility of the Engineer in Designing Technical Support Systems

The basis for supporting people in their everyday life is that the developer understands the user's needs and wishes as well as his or her individual living environment. To this end, it must be borne in mind that human beings can play two different roles in the evaluation and interpretation of products:

- In his role of a user, the human being takes a product and interprets its functionality through design and product-characteristics. By placing the functionality defined for himself in his life context, which is shaped by his individual life and action situation, the user decides on the usability of the product for himself. Accordingly, decisions for a product are not only characterized by its functionality but also by affective, emotional, and social aspects.
- In his role as an engineer, the human being takes up technological possibilities to implement these functions within a product in order to support the user and to expand his potential. This technology-driven mind-set is naturally oriented towards the needs of the user, but often reduces them to considerations of performance. Decisions related to the engineering point of view are characterized by physical connections on the one hand and, on the other hand, by rational aspects such as DFX criteria, technical and production feasibility.

Sarodnik describes these different views as *"mutual symmetric ignorance"* between user and engineer [Sar06]. Ultimately, it leads to the creation of functional and high-quality products, but these are not accepted by the user because they are not perceived by the user in the sense of problem solving.

The needs assessment for the planning and the conceptual design of products that are really intended to help people must not only take into account the life and action situation of people. A social responsibility of the engineer arises from the fact that products also have an assistance function [Gra17]. While performing everyday activities, people have to overcome multifaceted resistances. If the human being uses products to overcome his own shortcomings, this also works as training effect that is associated with a strengthening of competence and preservation. Otherwise, if products just focus on avoiding resistance in tackling the tasks, this can result in a loss of competence of the user [Gra17]. This results in the necessity to place the training effect in the product functionality above the purely compensatory functionality.

3 State of the Art

Three main approaches are common to describe users in product development: the methods of user participation, user experience (UX) and acceptance research.

User participation as an interdisciplinary concept encompasses a number of methods and approaches [Sar06]. Based on the definition of the development goals and the clarification of reasons for the integration of users in the development process (finding of ideas, product evaluation, and validation), the manner of integrating the user—from being a passive observation object up to being an independent innovator—has to be concretized. In addition, the target group must be defined. Decisions on the type of user integration that take into account the objectives lead to indications in which development stage the user's expertise is necessary. Summarized representations can be found in [Rei04, Fic05] for example. User participation sets the framework for integrating knowledge and expectations of users into the development process, thus supporting the transformation process. Difficulties arise from the used product models. The user can only partially access its functions, usually he receives explanations from the developer, because the models or prototypes are not intuitively interpretable. In addition, the test situation does not correspond to the real usage in everyday life. In total, this can lead to a falsification of results.

The methods of the UX research support the transformation process between user and developer. The aim here is to determine the usability of a product [DIN11], in particular affective, emotional, and psychological effects of its use. Hassenzahl refers to this as the "adventure of the user" [Has15]. UX research is not clearly delimited. Similar to user participation, UX comprises a set of methods for recording subjective aspects of product usage. Difficulties result from the absence of a human model on which the results can be evaluated. Ultimately, it is not clear which aspects or functions contribute to a positive perception of the product.

Last but not least, acceptance research provides numerous models that explain or predict the acceptance of products. A summary can be found, e.g., in [Bir14]. Based on the "Theory of Planned Behavior" many acceptance models were developed. According to Venkatesh [Ven00], the most important ones were integrated into the "Unified Theory of Acceptance and Use of Technology" (UTAUT). Acceptance is referred to as the behavior of the user when actually using the product. Common for acceptance models is that all relevant direct and indirect factors influencing the acceptance are known and considered. Nevertheless, it is not possible to make conclusions for the product resp. the functionality of the product. Thus, it is not possible to deduce how the product has to be changed in order to increase its acceptance, but predictions and assessments can be made with regard to its acceptance.

4   A Method for Describing the Everyday Practice of Human Beings

With the methodological approaches mentioned above, the significance of the product functionality for the user can be described, whereby the user is more or less taken into account as an individual only. However, these methods focus on product use. Here, a method for describing everyday practices is presented: The object of investigation is first of all routines of action in everyday life, and patterns in the conduct of life are to be recorded. Based on this, it is important to determine to what extent and where technical support is accepted and can be integrated in these daily routines. From this, it is necessary to derive functionalities, specifications for functionalities in the product as well as completely new ideas for technical assistance in everyday life.

Within the scope of a research project the everyday practices of elderly people in the home environment was investigated. A method for context-integrated, practice-oriented needs analysis (KPB methodology) has been developed, which consists of a set of qualitative survey methods and is based on sociological approaches to lifestyle and practice theory. This method is adaptable to a lot of other situations in daily life.

The everyday way of life is understood as an active achievement of a person, who is characterized by a high degree of habits and the spatial-material context. Everyday life does not take place "automatically", but is actively designed, whereby the process usually does not take place in a highly reflexive manner, but rather routinely and, as a rule, evades consciousness.

Another foundation is the theory of practice [Pon16]. From this perspective, the practical way of life is the central point for technological development, as it is the concrete "place" where support needs to be manifested. It is assumed that the use of technology should not "disturb" the familiar routines of everyday life. Something new is often encountered with a defensive attitude when it interrupts long practiced daily routines. From our point of view, this defensiveness should therefore be seen as a quasi "natural" reaction and not as a lack of willingness to innovate by the users. If one takes the practice theoretical argument seriously, then the direction of product development is given: systems should be developed which can be integrated into the existing lifestyle as easily as possible and which possess a high degree of practicality (see also [Bir16]). Then, there is a good chance that they will be accepted. In order to achieve this goal, day-to-day practices must first be identified and described.

For data collection, a set of qualitative survey methods has been composed. The first study focused on supporting the lifestyle of older people, since aspects of every-day routines are very pronounced here. 23 elderly, physically handicapped persons were interviewed twice. They are designated as research partners in the sense of a par-ticipatory research approach. The basic idea was to establish a triangular relationship between researchers, research partners, and the research topic "life management" in order to reflexively develop and analyze the latter in a joint dialogue.

The data collection was carried out in the domesticity of the research partners, which allowed the systematic inclusion of the material context. Verbal survey meth-ods such as interview, think-aloud method and reflexive dialogue methods were extended by elements of field research in the form of practical demonstrations. The first visit served to provide a comprehensive overall picture of the respective life situation. Essential parameters of living conditions such as material equipment, liv-ing environment, health situation, social integration, and education were asked for in a guideline-supported interview. At the same time, the initial interviews served to create a sustainable relationship of trust.

After the introduction, the focus was on the practices for dealing with everyday life. Since lifestyle is largely made up of routines, the main task was to make this to the object of conscious reflection; these should be evaluated by the research partners in terms of their difficulty in solving problems. Contrary to expectations, it was not easy to identify areas that seemed problematic from the subjective point of view of the elderly. As a rule, they had developed individual, sometimes highly creative handling strategies in order to cope with their limitations. Their competent "answers" sometimes concealed the underlying problems, which led us to systematically collect the handling strategies as well.

During the second visit, the focus was on handling strategies. Based on the trusting, equal work alliance, the elderly showed great willingness to demonstrate in practice how they deal with their everyday problems. This allowed us to understand problematic items in more detail, which formed the basis for finding ideas for technical solutions. Also, from the point of view of the research partners, the practices were not necessarily good solutions, since implementation was often associated with additional efforts and sometimes entailed considerable risks. In this respect, it can be expected that technical support aimed at these practices will have a great chance of being accepted. A detailed description of the methodology can be found in [Bir16].

5 Results and Implications for Product Development

5.1 Socio-scientific Results

A central finding was that we did not identify any problems, but always identified already "worked on"problem situations. The older people had cleverly and imaginatively developed strategies and practices to cope with the age-related limitations of everyday life. These practices were sometimes very simple and often not even visible at first glance. Nevertheless, they formed suitable "answers"to individual limitations often realised by using the simplest domestic inventories. Analytically, it is possible to differentiate between five different practices.

So-called *body techniques* are often used. Older people develop and establish, partly intentionally and partly unconsciously, physical handling routines in dealing with their challenges. One of the respondents had consciously developed a special body technique for climbing stairs. She entered the stairs diagonally and with both hands on the railing to slowly push her way up and down the stairs. If the potential of one's own body was not sufficient for the execution of everyday actions, its *enhancement* as a form of "technical upgrading" took place. This could be the walker, which allowed a person with limited mobility to cover distances, or the walker used as a means of transport. Another common technique was *empowerment*. Many older people consciously trained their existing skills to keep them stable. This could be the gymnastics in front of the TV set, but also climbing stairs or memory training on the computer.

And when the problems could no longer be overcome on their own, *social support* was actively organized, partly by their own children, but also by formal service providers who left their social environment untouched. We found changes in the material environment to be particularly important, especially because we often did not notice them at first glance. It was only gradually that many small, spatial adaptations were discernible, which the older people had to face in handling with their everyday life. An example: An elderly woman had "crammed" her hallway with furniture to hold onto them while walking. This "Furniture Walk" does not correspond to current considerations on accessibility, but can be an effective strategy to move forward.

5.2   Conclusions for the Engineering Perspective

Our quintessence from the qualitative survey: not problems, but the ways of dealing with them should form the basis for technical developments. What this means in concrete terms can be shown on the basis of the discussions on routines for action and a support hierarchy.

5.2.1   Importance of Action Routines

The focus of investigations lays on elderly people who live in their home environment and manage their everyday life largely independently. One of the most important findings was that their everyday life is determined by routines of action. Routines of action are defined as activities of everyday life that remain stable for a certain period of time and provide people with a framework for action [according to Has15]. They have generally grown over a long period of time. The fact that they are highly valued by the elderly can be attributed to the fact that they give structure to life and thus relieve the strain on action.

Strategies that older people use to deal with their everyday challenges and problems have also proved to be forms of action routines. These handling strategies can be tedious and involve considerable difficulties. Nevertheless, they are still capable of solving the respective problems, which means that they are no longer interpreted as problems by the elderly themselves. The development of routines is usually carried out creepingly, adapting to the restrictions that increase over time. In view of growing restrictions on mobility, an elderly person increasingly limits the amount of living space he or she uses by staying only in certain places, so-called "residential islands", which are easily accessible to them and which have been adapted to their needs.

In order to implement supporting technology systems in the everyday life of older people, it is necessary to adapt them to prevailing routines to make them being accepted. On the one hand, their ignorance casts doubt on the still existing competences of the elderly, what is usually seen as stigmatising. On the other hand, it does not correspond to the subjective problem definition of older people: if they perceive a problem as being overcome, their willingness to use technology to solve the problem will be small.

However, there is one exception: crises or incisive events such as a stay in a hospital, a move-out etc. lead to a break with routines that have been used up to then, with the consequence that in these situations new things—such as the use of technology—are more readily accepted.

5.2.2   Considerations for the Design of Technical Support Systems

The three-level support hierarchy outlined above [Pae12] has been confirmed and implies the following conclusions for technical systems.

Technical Systems for Training

At the first stage, technical systems can contribute to the independence of older people by motivating them to train their existing physical and cognitive abilities resp. to help practice these abilities. This level precedes practical everyday actions. This form of technical application corresponds to the empowerment (Sect. 4), which older people often choose as a conscious strategy to manage their everyday life even in the future. They want to maintain or strengthen the forces and abilities necessary for their routines of action. At this point, the engineer's knowledge of everyday routines is of secondary importance. The design of technical systems with regard to functionality can be relatively free, but the development of systems requires knowledge of competences and capabilities, their limits, and knowledge of the mechanisms by which these competences are formed. This appears to be possible only in close cooperation with somatically-centered departments such as medicine, gerontology, and sports methodology. Training support equipment should also take up aspects of the "Joy of Use" in order to generate positive success experiences.

Technical Systems to Support Everyday Actions

Technical systems can be used in a supportive manner on the second stage by assisting the implementation of problematic everyday practices and coping strategies. They can take over parts of everyday routines, make them easier or reduce the challenges of the material context. It is essential here to orientate oneself strictly to the daily routines, so that the technical system can be integrated without any significant effort and without the disruption of the routines. Only then an acceptance by the user can be expected. The functions of the technical system must always represent the action routines or parts of them. There are three forms of assistive technology.

Enhancement describes supporting systems worn directly on the body, such as hearing aids. These are connected to the body before performing the action. The user is supported during the action, but does not have to worry about the system. The product is only removed after the end of the action. According to the intensive interrelationship between technology and human, the product must be able to react to variations in the action routine resulting from the operating conditions—ideally without the user noticing this.

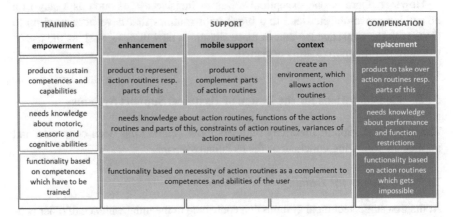

TRAINING	SUPPORT			COMPENSATION
empowerment	enhancement	mobile support	context	replacement
product to sustain competences and capabilities	product to represent action routines resp. parts of this	product to complement parts of action routines	create an environment, which allows action routines	product to take over action routines resp. parts of this
needs knowledge about motoric, sensoric and cognitive abilities	needs knowledge about action routines, functions of the actions routines and parts of this, constraints of action routines, variances of action routines			needs knowledge about performance and function restrictions
functionality based on competences which have to be trained	functionality based on necessity of action routines as a complement to competences and abilities of the user			functionality based on action routines which gets impossible

Fig. 1 Specification of the support hierarchy [Pae14]

Mobile devices describe systems that are not permanently connected to the user or the environment. One example is the rollator. Mobile supporting systems allow the user to gain more freedom, since they are only activated when required. This gives the user more freedom to use his or her own competences. The device is only used if the action is particularly strenuous. This can be associated with limited spatial or temporal availability.

Changes in context describe support systems that are firmly attached to the environment, such as handles in the bathroom. This means that the systems are also available for other users, but are less flexible and can only be used at the specific installation location. Such approaches appear to be effective in the home environment, but they also require an analysis of the routines of action.

Technical Systems to Compensate Lost Abilities

The use of compensatory technology on the third stage only becomes necessary when a requirement of everyday life can no longer be met even with support. But here too, the use of technology can contribute to enabling an independent life within one's own four walls. In this situation, the way of life has to be adapted accordingly, which means a deep cut in the user's lifestyle. The decision for compensatory technology is usually triggered by a crisis situation such as hospitalization or (further) illness. It is to be expected that technical systems that are considered in such situations will be accepted the better they can be integrated into remaining routines of action. Nevertheless, this form of technical support can be conceived relatively free from the routines of action. The decisive factor is the function to be fulfilled, which the user can no longer execute independently. These findings are illustrated in Fig. 1.

6   Conclusion

Technical systems can provide an important contribution to support people in their individual lifestyles and thus maintain their quality of life. However, the developer has a great responsibility in the development of technical systems to assist the user. It is not a matter of replacing skills, but of providing targeted support for actions in everyday life by means of technical systems. In this sense, the developer must understand the life and action situation and take this into account in the description of the target system and the requirements. The above contribution is intended to provide suggestions for this.

References

[Bir14]   Birken, T. (2014). IT-basierte Innovation als Implementations problem. Evolution und Grenzen des Technikakzeptanzmodell-Paradigmas, alternative Forschungsansätze und Anknüpfungspunkte für eine praxistheoretische Perspektive auf Innovationsprozesse, ISF. München.

[Bir16]   Birken, T., Pelizäus-Hoffmeister, H., & Schweiger. P: Technische Assistenzsysteme und ihre Konkurrenten: zur Bedeutung von Praktiken der Alltagsbewältigung für die Technikentwicklung. In VDE, Zukunft Lebensräume-Kongress 2016 (pp. 84–89). Berlin: VDE-Verlag.

[DIN11]   n.n. (2011). DIN EN ISO 9241-210:2011-01 Ergonomie der Mensch-System Interaktion – Teil 210: Prozess zur Gestaltung gebrauchstauglicher interaktiver Systeme. Beuth Verlag, Berlin.

[Fic05]   Fichter, K. (2005). Modelle der Nutzerintegration in den Innovationsprozess. Werkstattbericht Nr. 75, Institut für Zukunftsstudien und Technologiebewertung. Berlin.

[Gra17]   Gransche, B. (2017). Wir assistieren uns zu Tode. In P. Biniok and E. Lettkemann (Eds.): Assistive Gesellschaft. Multidisziplinäre Erkundungen zur Sozialform Assistenz. VS Verlag für Sozialwissenschaften. Springer Fachmedien Wiesbaden.

[Has15]   Hassenzahl, M. (2015). Experience design. Technology for all the right reasons. *Synthesis Lectures on Human-Centered Informatics*.

[Pae12]   Paetzold, K., & Wartzack, S. (2012). Challenges in the design of products for elderly people. In *9th International Workshop on Integrated Product Development, IPD Workshop*, Magdeburg.

[Pae14]   Paetzold, K., & Pelizäus-Hoffmeister, H. (2014). Structuring of application fields for technology for elderly people from the user perspective. In *International Conference on Human Behavior in Design*. October 14–17, 2014. Ascona.

[Pon16]   Pongratz, H., & Birken, T. (2016). Praktikanz als Zieldimension anwendungsorientierter Forschung. In Forum qualitativer Sozialforschung. Vol. 16; Nr. 3, Art. 9.

[Rei04]   Reinicke, T. (2004). Möglichkeiten und Grenzen der Nutzerintegration in der Produktentwicklung. Technische Universität Berlin, Dissertation Fakultät für Verkehrs- und Maschinensysteme. Berlin.

[Sar06]   Sarodnick, F., & Brau, H. (2006). Methoden der Usability Evaluation–Wissenschaftliche Grundlagen und praktische Anwendung, Bern: Huber.

[Ven00]   Venkatesh, V. (2000). Determinants of perceived ease of use: Integrating control, intrinsic motivation, and emotion into the technology acceptance model. *Information Systems Research, 11*, 342–365.

Acceptance Through Adaptation— The Human and Technology in the Philosophical and Scientific-Historical Context of "Sinnfälligkeit"

Kevin Liggieri

Abstract Fritz Giese—a famous scientist of psychotechnics—responded to a debate of man-machine-adaptation with his concept of object psychotechnics in order to adapt the environment to the worker. The idea behind this adjustment was, that man should be working in a trouble-free and decent environment, because of his complex biological nature, which allows him only to do productive work in the aforesaid environment. The economic movement or operation should always be intuitive and energy saving. The term for this problem was "Sinnfälligkeit", which was rather discovered by engineers than by psychologists. Below I will investigate in a philosophical and historical way how engineers and also psychologists operate with the concept of "Sinnfälligkeit".

1 Introduction

The special exhibition "Work Chair and Work Table", was organized by the German Society for Commercial Hygiene and the Reich Board of Trustees for Economic Efficiency at the German Museum for Occupational Safety on the 25th of May 1929 and then, because of its large appeal, it became a travelling exhibition through Germany (beginning in Southern Germany). Not only did it present examples of 'good' design but it also had an educational character for it was meant to bring the idea of an ergonomic workplace to workers and the wider public. The exhibition, which was the interdisciplinary work of industry doctors, industry supervisory officials and industrial engineers, had the aim of showing the "importance of a work economy" by a "purposeful formation of work table and work chair" [Ano31]. Premature fatigue or wear and tear should be prevented in this way, thus enabling more effective work.

K. Liggieri (✉)
Institute for Philosophy I, Ruhr University Bochum, Universitätsstraße 150, 44801 Bochum, Germany
e-mail: Kevin.Liggieri@rub.de

© Springer Nature Switzerland AG 2018   37
A. Karafillidis and R. Weidner (eds.), *Developing Support Technologies*, Biosystems & Biorobotics 23, https://doi.org/10.1007/978-3-030-01836-8_4

According to the ideas of the time, the human and the machine (workplace) were so incompatible that the attempt was made to somehow allow an adaptation of both sides to each other so as to let the overall system work efficiently. The exhibition was divided into two sections. The first section concerned the medico-physiologically correct posture, the second the connection to practice. In the first section, three life-size replicas of a seamstress were displayed, providing a view of the static work on the body, which illustrated fatigue through posture and hand-movements. A sitting skeleton should clearly signal to the visitors how work, through the right choice of support points in the body, can be alleviated and facilitated. In addition, the exhibition showcased a developmental history of the 'ergonomic workplace' by means of different work stools and chairs, each showing progress in relation to the previous model.

By this division between theory and practice the exhibition accordingly tried to didactically and reciprocally 'calibrate' the human and the work in the formation of the working environment. The physiologically correct posture at the work table were presented by models as well as by illustrated and static depictions for visualization and imitation. According to the contemporary Bauhaus architect Ludwig Hilbersheimer, the models presented were unfortunately "still purely theoretical", with only a few exceptions [Hil29]. Although prestigious companies like Siemens, AEG, Singer or Hinz & Stoll committed themselves directly to the new ergonomic chair, in 1929 the standardized work chair still remained an exception. According to Hilbersheimer, a "general usage" [would be] "extremely desirable" because every means which effected an "increase in production" without encumbering the working person should be used in practice [Hil29] (see illustrations in Figs. 1 and 2).

What this exhibition as well as Hilbersheimer's critical commentary make clear is the idea, which emerged ever more strongly in the 1920s, of the practical adaptation of the work environment (technology) to the human being. This adaptation refers on the one hand to the goal of disburdening the human being and thus effecting an increase in efficiency. However, on the other hand, it also shows how the view of the human changed. In contrast to Frederick W. Taylor's "Scientific Management",

 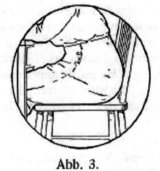

Fig. 1. Good and bad chair backs [Asc27, p. 97]

Fig. 2. Machine writers
with and without supports
[Asc29, p. 43]

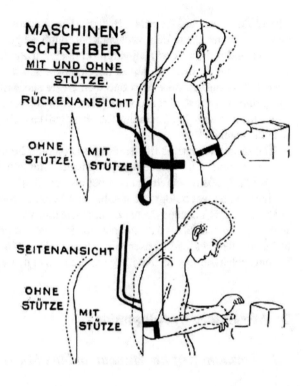

which in its studies of time and movement wanted to adapt the human being as an almost mechanical part to the work and the machinery, German engineering as well as in practical psychology—mainly in the branch that one called psychotechnics—increasingly addressed the human being as a 'factor' as well as a problem. The idea behind this adaptation was that, as a complex living being, the human being requires in its work a disturbance-free, orderly and ergonomic environment (including a work chair) in order to be able to work productively and efficiently. If one had hitherto tried to adapt the human being to technology through training and selection, one now also attempted to adapt technology to the human being. In short: if technology is to interact productively with the human being, then it inevitably had to be 'user friendly'—hence the humanistic sounding term 'human friendly'.

In what follows, this adaptation—which engineers nowadays call ergonomics, anthropotechnics or usability engineering—will be examined more closely, using the example of the "Sinnfälligkeit" in the object psychotechnics. The scientific-historical and philosophical approach is productive because it allows the discovery of fields of problems and discourses that are indeed argumentatively present today but which are by no means explicitly reflected upon. More and more emphasis is placed upon successful human-machine interaction in which trust and acceptance are conveyed by technology, complexity is reduced, and the human being is set at the centre (see the research program "Bringing Technology to the Human" of the German Federal Ministry of Education and Research, BMBF). The philosophical and scientific-historical questions connected to this "acceptance through adaptation" are thus: where do the motives and programs of adaptation originate? If the human being is to be taken as the measure in every technical development and if design and ethics should be measured against it, how does one then create successful (ethical and functional) acceptance in the use of technology? How does one generate the user-friendliness in which the human being (including the unskilled) can 'simply'—intuitively—interact with the machine?

The different motives of "Sinfälligkeit" should be discussed as forerunner and reference point of this modern development. "Sinnfälligkeit" is a concept from the 1920s and 1930s in which changes but also continuities in the interaction between the human and technology are notable in relation to contemporary discussions. An analysis of this concept makes comprehensible where the idea—so familiar and seemingly so sensible—that the machine is to be designed for the human being as an independent, individual user comes from and what image of the human being lies behind such assumptions. An image of the human that is still virulent today.

2   Acceptance by Adaptation

2.1   Formation of the Human and the Material

The psychotechnics of the psychologist Fritz Giese, the assistant to Walther Moedes (director of the Institute for Industrial Psychotechnics at the Technical University of Berlin Charlottenburg), is divided into two different functional categories. On the one hand, a technology that adapts the human being to the environment (machines, factories, social milieu), and on the other a technology that adapts this environment to the working human being. In the first case, as Giese reports in 1928 in his book "Psychotechnik", written for the general public, "the human, the subjective of the relation, [is calibrated] corresponding to the requirements of actuality" [Gie28, p. 8]. Here he speaks of "the subject psychotechnics", for example, in the form of aptitude tests or training procedures. In the second case, one can "proceed in a completely reverse fashion and bring the environment, the material into line with the natural psychological nature of the human; it is tailored to the relatively immutable nature of our specific character" [Gie28, p. 8]. Giese uses the concept "object psychotechnics" for the latter (regarding the adaptation of equipment, machines, lighting, advertising material to the psycho-physical preconditions of the human). Consequently, the subject psychotechnics deals with training, professional issues, professional advice and treatment of the human being (understood as a psychologically influencing of the human with the goal of improving performance). In a certain sense object psychotechnics can be read as analogous to "Sinnfälligkeit". The human and its 'nature' here become the measure of things. Space and equipment must be directed according to it. The worker and its statutes will now, from the perspective of the psycho-technicians and engineers, no longer be ignored and no longer simply embedded in mechanical analogies but will become the special point of access for investigations that adopt the 'human measure'. This is what "Sinnfälligkeit" was all about.

There is already a specific evocation of acceptance in this approach to the human being because human-calibrated machines not only provide greater security but in effect also trust. In spite of the humanistic rhetoric of many authors, the human nonetheless remains economically a 'matter' which had to be carefully considered because calculations and consequences were harder to measure in its dynamic inter-action with the environment. The main problem here was to establish an order in the extraordinary 'human' factor, which cannot be rationalized so easily due to its embod-iment. This is precisely why the human being became conceived as an important part of production: through its inattentiveness it might also destroy other parts—espe-cially the expensive machines—and thus can cause great financial damage. From the economic perspective, the human being and machine were accordingly materials to be adapted to one another for the higher purpose of optimal organization in the con-cept of object psychotechnics/"Sinnfälligkeit"—as previously in the US-American Taylorism.

In a second step, a closer look at the object psychotechnical formation of work equipment and workplaces reveals the narrow relationship between work psychol-ogy, physiology, and psychotechnics and, with this, the differentiated treatment of a human 'psychophysics'. A focus on the human being centered on anthropological peculiarities is inevitably opened up by looking at human errors, accidents and per-formance constraints. What is important is that this biological entity, the living being in itself, was never purely rational, mechanical or uniform but dynamic and partly unpredictable. How, according to the question of the engineers and psychotechni-cians, could one adjust work equipment to this 'living' being and thus generate acceptance?

2.2   On the 'Sinnfälligkeit' of the Working Equipment and the Workplace

Through an efficient working environment (here: workplace and equipment) adapted to the psycho-physical peculiarities of the human, the economic benefits of work should increase. Fabian consistently uses Giese's concept of an "object psychotech-nics" for the "adaptation of the object" to the "mental nature of the human being" [Fab30, p. 621]. According to Fabian, the heterogeneous region of object psychotech-nics (accident prevention, advertising material, work equipment, lighting) can be divided into motion studies in the factory and the rationalization of the workplace. The motion studies were, on the one hand, still very much oriented to Taylor's studies of time and movement, on the other, however, they were also directed to studies of fatigue as well as to the effects accompanying work-technical circumstances. The second field of a, rationalization of the workplace' required, in contrast to the general studies of the first point, exact special research, that is, field research in a specific workplace. What was at issue in this on site research was no longer general reflections but an exact, analysis of the workplace' and its environment (heat, air, light) [Fab30, p. 627]. The psychologists, who were to be instructed on specialist questions by technicians, examined various factors such as material, work tools and machines in their peculiarities, types and usages, always with a view to the psychophysical impact on human beings.

This optimal, economic movement or operation should thereby always be 'natural' and thus intuitive and energy saving. The efficacious term for this was "Sinnfälligkeit" which, as Giese himself remarked, was discovered by engineers rather than by psychologists. "Sinnfälligkeit" thus means a literal "falling into sense", a commonsense obviousness, of the right correlation. This 'falling' should happen unconsciously and disburden, to speak with the psycho-technicians, reflective consciousness.

The operation should therefore be self-explanatory in the best case, like the touch screen on today's mobile phones. One acts appropriately but does not have to think about it. Since the machine can be operated intuitively without direct knowledge about it, one does not need either well-educated workers or overly skilled operators. The technology 'explains' itself through its human-centered design. On the other hand, however, the psychologists and engineers were able to make use of this 'falling into sense' in advertising and operational elements, because through constructions not only bodily but also mental performance (and along with it mistakes, accidents and loss of time) were reduced and intuitive dexterity could be increased. According to Giese, 'falling into sense' thus related to different forms like traditional associations, habits, physiological disburdening of movement or instinctive movement.

This mediation proceeded not only from an instinctive use of technology but equally from a very specific anthropophilic technical design. The fundamental idea that appears in discursive and practical dealings is: the technology should and can support "reactive moments" by suitable design [Fab30, p. 660]. To put it simply: technology is not the human being's opponent, but its partner. In turn, the human was no longer simply a mechanical motor in which it was sufficient to analyze the physiological course of movement but, in its connection with the machine, it was a sensory motoric component whose psychical peculiarities (such as attentiveness, perception, dexterity) were brought to the fore. In this emphatically charged argumentation, the simple and intuitive handling of technology did not only create security and trust but also acceptance. The machine is no longer perceived as a threat when it can be operated securely and intuitively. In the best case, the machine (for example, the car) is perfectly adapted to the human being and its movements and senses. It is phenomenologically fused with the driver as a technological device. The driver can simply let themselves 'fall into the sense', (i.e., because of bad construction) hence they fall out—an accident occurs. Trust, safety, and acceptance are therefore strongly dependent on the adaptation of the technology to the human being. It became quickly clear to the psychotechnicians, who in such cases looked more closely at the manipulations of the equipment, that the handbrake must be pulled "backwards". For one thing, for reasons of the expenditure of force (that was already clear in the experiments on the position of the grip) and for another—and this seems epistemologically interesting—for reasons concerning tradition in the drawing up of horses. This mediation of old media into the new at the center of successful adaptation is still evident in the process of design and production today. Hence the Windows Desktop has a "recycle bin" and a "typing page", i.e., pre-digital artefacts, that resurface in a mediating role in the new medium for a successful acceptance and 'falling into sense' or "Sinnfälligkeit". The user sees something familiar in the strange. The 'uncanniness' of the unfamiliar medium recedes. The human being and its senses as well as its cognitive capacities do not have to adapt themselves to something wholly new in good "Sinnfälligkeit" (although the PC, unlike the typewriter, represented something precisely so revolutionary an innovation on the level of software and hardware) but finds a communicative connection. Giese's inferences by analogy were thus used to shape the course of movement more efficiently, whereby he left unconsidered that such well-known instances 'falling into sense' are always culturally and even generationally bound.

To cite a further example, we need only refer to the associative problems that would arise in a simple visit to someone's house if the lowest doorbell did not connect to the ground floor and the highest doorbell to the highest storey. These kinds of traditional associations enable and generate a capacity for connections. Acceptance is thus achieved by the annexation of well-known and thus already internalized ways of acting. Complementary to haptic adaptation to machines, visual identifications of work material such as red color as a warning or flashing as a sign of danger were in the engineering-psychological context of the 1920s examples of an "obviousness" and "mnemonic aid" for the "simple" worker [Gie27, p. 600]. The psychologist Hans Rupp spoke in this sense with the terminology of Johann Friedrich Herbart's of the "narrowness of consciousness" which, on the one hand, was characterized by a restrictedness of the human receptive capacity and on the other was characterized by a dependence of the order and structure within which signals and impressions reach the human being [Rup28, pp. 18–19]. The goal thus lay in delivering signal, symbol, and significance in a "meaningful context" [Gan31, p. 251]. In a similar way to contemporary technical design, what was striven for was a close, unreflecting overlap between the sign and meaningful content, whereby the human being quickly and intuitively recognizes its mandate for action and does not perceive the interface itself as a barrier or a problem. A last example: the "falling into sense of movement", that is made efficacious here is also reflected imperatively in the "hand rule" that underlaid and still underlies most technical designs: "shifting and turning to the right leads right, moving in (opens) or increases. Shifting or respectively turning to the left leads left, moving out (closes) or diminishes" [Gie27, p. 603].

In most of the demands for Sinnfälligkeit not only the body and its movements but thinking too should be disburdened and not 'stand in the way' of the user. Through the appropriate design of the user interface, the human being should be able to quickly, safely, and efficiently arrive at its chosen goal without obstructions and losses (be these human or mechanical). Human activity thus becomes an intuitive and thus less fallible reflex.

3 The Machine Is not Your Friend

When one looks at the explanations as well as at the question initially introduced concerning acceptance through adaptation in the 1920s with the advent of Sinnfälligkeit and object psychotechnics, the view of human-machine interaction seems to change although different paradigms (familiarity, simple operation, universal user, accident prevention) have remained constant since then to this day. Where the human being in studies of time and movement of 1920s Taylorism represented an error factor, psychotechnics saw in it psychological and physiological possibilities as well as limitations. These possibilities and limitations also reappear in the modern view, for example that of the German Federal Ministry of Education and Research BMBF. Its 2015 research program—with the telling title "Bringing Technology to the Human"—clearly focuses on a human-machine interaction that (with a view to political, social, and financial support) is not exactly subtly positioned as anthropophilic. The program focuses on a "hand in hand" relationship between the human being and technology [BMB15, p. 5], in which a "responsible" technology [BMB15, p. 6] not only serves human beings and thus moves them to the "center point" [BMB15, p. 7] but shall also increase "acceptance" as well as "trust" by "ethical reflection" and design [BMB15, p. 6]. These motives were, as already shown, already argumentatively as well as technically laid out in "Sinnfälligkeit" and have been differentiated through information-processing machines. However, the humanistic maxim of a 'cooperation' based on partnership between the human being and the machine as well as a human-centered operation has remained the same.

According to the credo of "Sinnfälligkeit" as well as of modern usability engineering, what is combined in the human factor is, on the one hand, the fear of unpredictable and surprising irrationality in the form of performance fluctuations and, on the other, an opportunity for the ethical aspects of work (humanization). On this interpretation, the human being is indeed in a certain sense subordinate to the machine, on the other hand the machine must be ethically 'adjusted' to it. It was apparent in the present analysis that a natural-scientific exactitude cannot be simply transferred to the human being and its statutes. In the descriptions then and today, the living being seems too complex, too chaotic, too incalculable for a simple mathematization. It was thus clear to many of the researchers in psychotechnics that the problem factor 'human being' required a new treatment of its own. This was provided by the concept of "Sinnfälligkeit". The opportunity therein was that of seeing the human being as a human being and no longer as a slavish muscle 'machine'. The task was to adapt the machine such that it 'falls into sense' for the living human being.

In this sense, not only the machine was changed in the argumentative and constructive process of "Sinnfälligkeit", but also the human being. However, it was not merely a humanistic aspect that was associated with this but rather a more economic one: in the 'human friendly' and 'user accessible' machine, it was not unconditionally better for the human being, but its work was more efficient, more disturbance free and thus more productive when its interaction with the machine 'fell into sense'. The fact that these economically dominated, as well as disciplining, accesses to the human being are also still present in the modern forms of an ergonomic usability engineering and product design, should at least be noticed—better yet, understood—before one speaks unreflectingly and prematurely of a 'humanization of work' or of 'human-friendly' design. The machine is, thus spoken, not your friend but at the most your work colleague and communication partner. The "Sinnfälligkeit" (in modern terms: usability) of the interface, however, contributes to the machine being no longer regarded as something other, unknown and non-human.

References

[Ano31]   Anonym. (1931). Wanderausstellung Arbeitssitz und Arbeitstisch. In Hygienischer Wegweiser 6, p. 178. See also Alexander, J. K. (2008). *The Mantra of Efficiency: From Waterwheel to Social Control*, (125, pp. 101–125). Baltimore.

[Asc27]   Ascher, L. (1937). Sitze und Tische in Gross- und Kleingewerben. In: Zentralblatt für Gewerbehygiene (4, pp. 97–100).

[Asc29]   Ascher, L. (1929). Zweckmäßige Gestaltung von Arbeitstisch und -stuhl. In: Psychotech-nische Zeitschrift (2, pp. 43–45).

[BMB15]   BMBF (2015). Technik zum Menschen bringen. Forschungsprogramm zur Mensch-Techik-Interaktion. Bonn: Thiel.

[Fab30]   Fabian, G. (1930). Einführung in die Psychotechnik des Arbeitsgerätes und des Arbeitsplatzes. In F. Baumgarten and G. Fabian: Psychotechnik der Menschenwirtschaft. In F. Giese (Ed.). Objektpsychotechnik. Handbuch sachpsychologischer Arbeitsgestaltung, (pp. 619–684). Halle a.: Carl Marhold.

[Gan31]   Ganzenhuber, E. (1931). Normung und Psychotechnik. In Maschinenbau/Der Betrieb, (10, pp. 250–252).

[Gie27]   Giese, F. (1927). Methoden der Wirtschaftspsychologie. In E. Abderhalden (Ed.). Handbuch der biologischen Arbeitsmethoden, Abteilung VI, Teil C, Band 2. Berlin/Wien: Urban und Schwarzenberg, pp. 119–744.

[Gie28]   Giese, F. (2017). Psychotechnik. Breslau: Ferdinand Hirt (1928). See for this, Liggieri, K. "Sinnfälligkeit der Bewegung"—Zur objektpsychotechnischen Anpassung der Arbeitsgeräte an den Menschen, in Zeitschrift für Technikgeschichte (1, pp. 29–62).

[Hil29]   Hilbersheimer, L. (1929). Arbeitssitz und Arbeitstisch. In Die Form: Zeitschrift für gestaltende Arbeit (4, p. 362).

[Rup28]   Rupp, H. (1928). Die Aufgaben der psychotechnischen Arbeits-Rationalisierung. Iin Psychotechnische Zeitschrift 6, pp. 17–19.

Biomechanical Analysis: Adapting to Users' Physiological Preconditions and Demands

Andreas Argubi-Wollesen and Robert Weidner

Abstract Exoskeletal systems for the workplace are mostly designed to reduce strain and to prevent musculoskeletal disorders. In order to design these systems accordingly, biomechanical and physiological demands of the workplace and the individual's response to these demands have to be known. Hence, biomechanical aspects during application of the exoskeletal systems have to be evaluated. Biomechanical analysis delivers tools and methods to investigate responses of users caused by the interaction between user, workplace, and exoskeleton. This section summarizes common methods for investigations on body movement, muscular and metabolic activity, applied forces, and soft tissue constraints.

1 Introduction

In recent years, there has been a steady rise of physical support systems such as exoskeletons. The claim made by developers of these systems is, that they can be beneficial in the reduction of work related musculoskeletal strain on the human body [Loo16]. Therefore, the development of support systems for specific work scenarios not only comes with technological challenges but also with the need for a deep understanding of

(a) the amount and characteristics of load induced by the workplace scenario,
(b) the way humans tend to react and adapt to these workloads,
(c) the net-influence of physical support systems as well as
(d) the resulting biomechanical effects caused by the use of physical support systems.

A. Argubi-Wollesen (✉) · R. Weidner
Laboratory of Manufacturing Technology, Helmut Schmidt University/University of the Federal Armed Forces Hamburg, Holstenhofweg 85, 22043 Hamburg, Germany
e-mail: argubi-wollesen@hsu-hh.de

R. Weidner
Chair of Production Technology, University of Innsbruck, Innsbruck, Austria
e-mail: robert.weidner@hsu-hh.de

© Springer Nature Switzerland AG 2018

A. Karafillidis and R. Weidner (eds.), *Developing Support Technologies*, Biosystems & Biorobotics 23, https://doi.org/10.1007/978-3-030-01836-8_5

If the specific demands of certain workplaces are not fully understood, exoskeletal systems might not be able to support the user in these scenarios to their full potential. Worse, they might even present themselves as an additional load to the user, contradicting their intentional value for the user in the first place [Ras14]. Because of this, developers of exoskeletal systems for specific use cases profit from a biomechanical and physiological analysis with regard to at least two aspects:

(a) The specific demands of the use case or workplace. Overall, these can be described as the total work load (weights which have to be carried, duration of the work, amount of repetitions per time interval, etc.).
(b) The individual physiological strain due to the work load [Bec17].

Based on these findings, areas in need of support can be identified and potential solutions or enhancements via the application of physical support systems can be depicted. These are the fundamentals, on which physical support systems have to be developed in order to be well adapted to workplace demands. Throughout the development and especially during implementation phases, these systems have to be thoroughly tested for their ability to actually deliver on their promises—above all, reducing the individual physiological strain.

2   Consideration of User Heterogeneity

Developers of physical support systems for specific workplaces face a serious challenge. Although certain workplaces come with an objective workload such as specific weights to handle as well as a pre-defined range of handling manoeuvers per time interval, physiological responses to these loads are highly individualized and depend, moreover, on the used support system. Differences in, e.g., gender, height, weight, and strength level will create a very heterogeneous response to the actual workload inside the workforce. In the foreseeable future, exoskeletons will be based on a general design, with only small amounts of individual adjustments. Therefore, the basic design elements must be able to address a wide range of potential users, leading to the necessity of systematically testing their effects.

3   Evaluation of Biomechanical Effects

In general, what has to be compared is the unsupported condition (work task without support system) with the supported condition (work task with support system) and see how user responses to these conditions will differ in order to evaluate the effects of the support system on the user. Due to the individual responses, it is necessary to include a variety of different users into the analysis, especially if a statistical sound analysis of effects on a wider population is to be achieved. It is recommended to compute an a priori power analysis before conducting a biomechanical study. The a priori power analysis determines the amount of test subjects (sample size N) needed if a certain statistical power for a given level α is to be achieved [Fau09].

In what follows, different methods and technological approaches of biomechanical analysis are presented, covering a broad range of the current tool-set used in biomechanics for the evaluation of physical support systems.

Each method presented here is only one, but certainly not the only way, of how certain biomechanical aspects of the interaction between workplace, human, and a technical system like an exoskeleton can be investigated. After a short explanation of the general aims of each method and its general technological approach, the challenges of adapting the method to the requirements of biomechanical analysis of physical support systems are briefly discussed.

4   A Selection of Relevant Biomechanical Methods

Biomechanical analysis mostly comprises different aspects of a human's interaction with his environment: the motion itself (kinematics), outer forces (kinetics) applied to the body or caused by its interaction with the environment and inner forces as well as muscle activity which cause voluntary body motion. Additional to the biomechanical data acquisitions, cardiopulmonary data such as heart rate, respiration gases etc. are often times measured to incorporate physiological responses to workload and possible benefits of exoskeletons in these regards. The following table summarizes a selection of methods used for evaluation of biomechanical effects including outcome parameters (Table 1).

4.1   Motion Analysis (Kinematics)

The use of physical support systems will indeed change human motion patterns at least somewhat, as wearing a device strapped around human joints always increases weights to the limbs or alters the degrees of freedom for the joints at least partially. Thus, it is of the utmost importance, that these changes will not be as severe as to create unnatural or even non-ergonomic motion patterns, thereby contradicting the intent of using the system as a means of reducing physical strain on the human body. A biomechanical approach of looking at human movement is by way of analyzing the body kinematics of the skeletal system which can be described in terms of position and orientation of body parts, velocities and acceleration of, e.g., body segments or joints or simply in joint angles between adjacent body segments (motion capturing). In order to test for changes in motion patterns induced by the use of wearable support systems, this kinematic data has to be compared between a baseline (work task without the support systems) and the same task while wearing the support system. Historically, kinematics has been evaluated by analyzing two-dimensional video data, resulting in reporting two-dimensional joint-angles and (angular) velocities and

Table 1 Selection of biomechanical assessment tools in the evaluation of physical support systems

Object of investigation	Biomechanical method/assessment	Outcome parameters (selection)
Body movement/motion	3D-kinemetry (optical systems or inertial measurement units)	Characteristic motion patterns [joint angles, trajectories, dynamics (velocities, acceleration)] with and without support
Muscle activity	Electromyography (EMG)	Muscle activity in percent of maximum muscle activity (%MVC, maximum voluntary contraction), frequency spectrum of the EMG-spectrum
Cardiovascular/cardiopulmonary activity/metabolic effort	Spiroergometry/heart rate measurement	Heart rate, O_2/CO_2 relation, relative VO_2 in ml/min/kg, etc
Soft tissue constriction	Near-infrared spectroscopy, questionnaires	Oxygenation of the blood, changes in blood flow inside the human soft tissue
Force/torque (internal, external)	External forces (dynamometry), internal forces (inverse dynamics)	Ground reaction forces, joint torques, instabilities, and balance (body sway movement of center of pressure), changes in force distribution (time/location)
Individual perceptions of effort, strain, comfort etc./health status	Questionnaires	Health status, user acceptance, comfort level, level of perceived exertion, confounding factors, etc

accelerations. The value of the use of two-dimensional joint angles is the transparency for most readers including laymen as, e.g., a reported two-dimensional 90° angle between upper arm and forearm is something most people can relate to. Most people will automatically assume a plane which is spanned by the upper and the forearm acting as an x and y vector, regardless of the orientation of both segments in the room. However, two-dimensional video data can only be used if the video camera is pointed perpendicular to the imaginary plane. As almost all work tasks involve circular motions or rotations around the longitudinal axis, two-dimensional video data is insufficient for motion analysis. To overcome these shortcomings of 2D-Video data, the use of three-dimensional motion capture systems and methods has become commonplace throughout the last two decades. Nevertheless, in reporting kinematic data such as joint angles, it is still common to use two-dimensional values. Most 3D-kinematic toolsets can be divided into two distinct technical approaches: optical systems and a networks of inertial measurement units.

4.1.1 Optical Systems

The use of optical systems is widespread in biomechanical labs throughout the world and still serve as the gold standard when it comes to precision and reliability of kinematic data. Although there are different systems on the market, the basic principle of optical systems is the use of a multitude of infrared cameras around the subject to be analyzed (a minimum of six to eight cameras mostly). Optical motion capture systems will track either one of two different types of markers attached to certain pre-defined areas of each of the subjects' body parts or segments: active markers emitting infrared light or passive markers, reflecting infrared light beamed upon them via infrared light emitting cameras.

If the infrared light (emitted or reflected) is picked up and received by two or more infrared cameras, the optical system can compute the three-dimensional positional data of these markers inside the observed room (capture volume) by triangulation. As a result, the markers represent the positional data of the attached body segments in a three-dimensional volume. In order to create kinematic data out of this information, an underlying biomechanical model is used, which will, e.g., calculate joint angles or angular velocities based on the data. Depending on the topic of the investigation, different marker-sets are being used. Two distinctive types of marker-sets can be distinguished: anatomical and cluster marker-sets.

The first type makes use of palpable bony landmarks where markers are attached to, the latter uses pre-defined sets of multiple markers per body segment. Both types do come with specific advantages and disadvantages. What is to be considered is, that even though optical motion capturing methods are commonplace in biomechanical analysis, there is no standardization for certain marker-sets nor for the mathematical calculation of joint angles etc. Therefore, prior to any investigation, the marker-sets as well as the underlying kinematic calculations in use should be assessed regarding validity and reliability. A systematic review about different approaches can be found in [Val18].

A major drawback in the use of optical systems for the evaluation of wearable support systems is the potential occlusion of the markers by the support system. Depending on the size and design of the support system even the attachment of markers to certain body parts in itself can be problematic [Du16]. Furthermore, as the calculation of joint angles etc. are mostly based on pre-defined marker-setups, changing these setups due to restrictions presented by the support system will result in additional validation efforts for the newly created marker-sets, see exemplary in Fig. 1.

4.1.2 Inertial Measurement Units

Another technological approach to assess body kinematics without the need of an array of mostly highly expensive infrared cameras as with optical systems, is the use of inertial measurement units. As there is a whole paper on the use of inertial

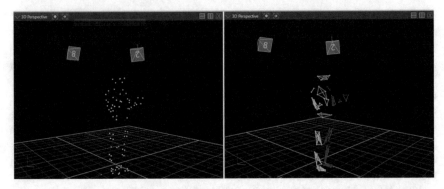

Fig. 1 A digitized subject with an optical marker-set of anatomical and cluster markers (left), body segments are defined by clusters (right) in Vicon Nexus^{TM}

measurement units (IMUs) for motion capturing included in this book, the following explanation will be very short, and readers are kindly referred to part III.

In contrast to optical systems, the IMU-System does not rely on cameras and markers, but on inertial measurement units attached to body segments. Normally, if joint angles are to be calculated, the use of a minimum of one IMU per adjacent body segment is necessary. Therefore, the ability to record kinematic data for the whole body will result in the need of ca. 20–25 IMUs, depending on the biomechanical model in use. The IMUs will detect changes in space coordinates, velocities, or the magnetic field. The biomechanical model will provide certain assumptions of how the IMUs can move in conjunction in form of model-based sensor fusion algorithms. Based on these, the kinematic data can be calculated.

Even though the potential optical occlusions of the IMUs by a wearable support system are of no consequences, unlike with the optical systems, the use of IMUs in this context is not without challenges. As with the marker-sets of the optical systems, the biomechanical models in use do make certain assumptions on where the IMUs are placed on the human body. Hence, depending on the support system, these areas might be unavailable or the sensors will hinder the range of motion of the user while wearing the support system. In addition, certain IMUs do react very sensitive to any electrical system close by, resulting in unreliable or disturbed data. Especially with active support systems, the use of IMUs for biomechanical data present their own challenges due to these constraints. At least, the models in use should be validated against another reference, if possible against optical systems [Rob17].

4.2 Analysis of Forces (Kinetics)

The use of kinetics in biomechanical analysis for the evaluation of the effects of exoskeletal systems can be separated into two distinctive parts: the determination of external forces caused by the subject during interaction with the exoskeleton and the calculation of internal forces and torque generated in the joints [Alk01].

External forces caused by the user (with and without the use of an exoskeleton) are mainly assessed with force plates or force detectors. Force plates, using strain gauges or more commonly piezoelectric sensors, are mainly used to determine the amount and direction of ground reaction forces. The assessment of forces generated by the user can be helpful in answering questions of how the use of physical support systems will, e.g., alter the user's stance and how forces are projected throughout the body [Sin09]. There is a longstanding tradition in gait analysis regarding the utilization of force plates to assess ground reaction forces and other outcome parameters derived from this information.

Force plates are mostly used in the biomechanical evaluation of exoskeletons for the lower legs (or the whole body for that matter), aiming for assistance during gait. As force plates are an integral part of clinical gait analysis, it makes perfect sense. Force plates can detect the amount and direction of ground reaction forces during contact phases of the gait. Therefore, any differences in kinetic-temporal parameters between a supported and unsupported gait are clear indications of effects caused by the exoskeleton. Exoskeletons with a focus on rehabilitation purposes aim to change these parameters in directions of those seen with healthy subjects (knowing very well that gait patterns are highly individual).

The use of force plates at the ground is also beneficial in the evaluation of upper limb exoskeletons, even when there is no walking involved, e.g., at stationary work places, as exoskeletal systems worn on the upper body influence postural stability and body sway. A human standing has to maintain its posture by generating muscle forces around the joints (ankle, knee and pelvic) that will counter body sway that will lead to postural instability. The human body acts similar to an inverted pendulum in this regard, with constant motion around its center of gravity. This effect (the constant body sway maintained by muscular activity) is easily felt when closing the eyes while standing upright.

Upper limb exoskeletons, especially active ones with carried power source at the back, have a similar effect on the maintenance of postural control as regular backpacks do. As these cause compensatory kinematic changes at the trunk (forward lean) or at the pelvic (pelvic tilt) and increase body sway (higher loads cause higher sway velocities [Str17]), it is safe to assume that exoskeletons for the upper body will cause similar effects. Additional to the weight of the exoskeleton, any of the aforementioned effects on the human body will add load on the user, thereby acting detrimental towards the exoskeletons aim of reducing physiological strain. By analyzing the ground reaction forces during work tasks while wearing a support system (in most cases, a test subject will stand on a single force plate), the amount of additional load forced upon the body by the support system is verifiable and its effects on body position and disturbances of posture. Any changes to the posture will cause differences in the projection of the center of pressure (COP) on the surface of support (the ground).

Furthermore, with the help of measured ground reaction forces in combination with kinematic data, the internal net joint moments can be calculated using inverse

Fig. 2 Reduction in muscle activity compared between baseline and two supported scenarios with an exoskeleton [Ott18]

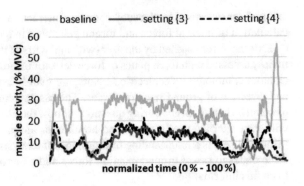

dynamic equations [Fra14]. Comparing these between supported and unsupported tasks will deliver another layer of expertise about how the exoskeleton will alter joint kinetics [Hwa15] and if its use is to be considered beneficial or potentially harmful to the user [McG17].

4.3 Analysis of Muscle Activity (Electromyography)

Maybe the most used biomechanical analysis in the evaluation of physical support systems of all presented methods in this section is the electromyographical analysis (EMG) [Cho13]. What can be derived from EMG-Analysis is the electrical activity of the muscles' motor units. However, it is necessary to address some misconceptions regarding EMG-analysis being quite widespread in non-biomechanical domains. Due to the stochastic nature of the EMG-signal, it is *not* a direct measurement of strength, especially not in comparison to other people. Furthermore, there is no directly measurable relationship between the amplitude of an electromyographical signal and the exerted force by the user [Hal12].

In EMG-Analysis mostly non-invasive surface EMG-sensors are placed upon the skin (surface EMG). In order to evaluate the EMG-Signal it has to be normalized against a reference value. This reference value is created by maximum voluntary contraction (MVC) measurements directly prior to the actual measurements. The general idea behind MVC measurements is that these will present themselves as a 100% contraction value. Any of the actual EMG-measurements will be presented as a percentage value in reference to the MVC [Ott18], as shown in Fig. 2. An important distinction is, that these MVC measurements are mostly done with maximal isometric contraction, that is, the EMG-Signals derived from these, can be even higher than the MVCs which is due to the highly dynamic movements during the actual measurements [Hal12]. This makes the 100% assumption of the MVC kind of misleading. Granted, most work tasks involve slower movements so that these shortcomings of the MVC measurements do not play such a big role in these cases as, for example, compared to highly dynamic movements in sports.

One of the biggest hurdles in maintaining a high-quality signal, which can be further analyzed, is the reduction of various signal noise or artifacts. Noise can be based on different internal or external factors. An internal factor is the amount of fat tissue under the skin, causing changes to the impedance between sensor and motor units. External factors such as electromagnetic noise produced by running machinery or the support system itself will also affect the signal negatively [Bux12]. The latter is one of the main reasons why sometimes EMG-measurements at the work place will fail. Although EMG-measurements are of high value for biomechanical analysis at the work place due to the small amount of technical equipment needed, especially in industrial facilities, the amount of electromagnetic noise produced by heavy machinery will hinder EMG-measurements. In cases where the electromagnetic noise will be only induced by power outlets, this so-called Power-Line-Interference, occurring mainly in the 50 or 60 Hz spectrum, can be removed by digital filtering afterwards without taking away too much of the signal's information.

Furthermore, the EMG-signal is highly sensitive towards motion artifacts [Bux12]. Any external disturbance of the spatial relationship between sensor and motor units of the muscle will alter the signal quality tremendously and any post-measurement filtering method will reduce the signal's quality even further. Unfortunately, many physical support systems such as exoskeletons will indeed create motion artifacts through pressure on top of the sensors in places where the exoskeleton will cover body areas which are of interest for the evaluation [Ruk16].

Therefore, it is of high importance to create a test scenario, where most of the mentioned disturbances, whether through external influences or through motion artifacts, are taken care of in the first place.

4.4   Near Infrared Spectroscopy (NIRS)

The use of physical support systems always ensues the challenge of strapping the system to the user in one way or another [Jar12]. Doing so without harming the user is contingent on holding any system-human interface in place without using straps that dig into user's soft tissue resulting in insufficient blood flow and skin irritations.

A novel approach of testing for any insufficient blood flow induced by the support system is the use of near infrared spectroscopy (NIRS) [Ham11]. NIRS is a non-invasive optical measurement tool. Its main purpose is the identification of concentration changes of oxyhemoglobin (Hb0) and de-oxyhemoglobin in the blood. It is primarily used in cortical signal analysis such as in visualization of brain activity, leading to its use as the tool of choice for brain-computer interfaces. Nevertheless, the possibility of measuring blood flow as a secondary outcome value of NIRS measurements [Jon16] makes it a valuable tool for measuring restricted blood flow as a result of compressed or constricted human soft tissue at human-exoskeleton interfaces (see Fig. 3). The amount of reduction in blood flow could then potentially be used as an indicator for the severity of soft tissue irritation induced by the human-exoskeleton interface. Given the changes in interaction between interface and human

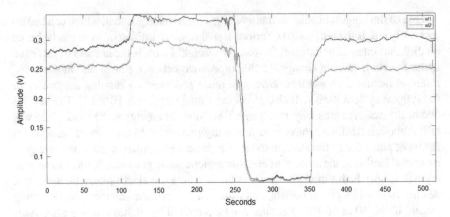

Fig. 3 Example of restricted bloodflow due to local muscle compression between 250 and 350 s, measured with a NIRS system (upper line represents oxygenated and lower line deoxygenated blood)

soft tissue during the actual motions while wearing the exoskeleton like changes in pressure, relative movements between skin and interface [Cem14], etc., such analyses are by no means easy to conduct or to interpret. Currently, there are only few biomechanical studies which make use of NIRS as a way of describing the amount of occlusion of blood flow let alone using it as a tool for analyzing negative side effects of human-exoskeleton interfaces.

Many users of exoskeletons report a serious restriction in freedom of motion and discomfort. Assuming the exoskeleton provides sufficient degrees of freedom for the supported joint, it must be the interface which needs to be optimized. If NIRS does present itself as a viable and cost-effective method to measure the strain on the soft tissue, it is worthwhile to take a closer look at this method in future biomechanical research when it comes to ascertaining the effects of exoskeletons on its users.

Most of the aforementioned biomechanical measurements take a look at very specific body parts. The EMG data yields insights concerning the muscle activity of local muscle areas and the kinematic data is mostly interpreted on the basis of joint angles of adjacent body segments. What is missing is the use of global parameters, which can express the amount of strain reduction caused by the use of an exoskeleton.

It is quite safe to assume that the use of a specific exoskeleton might reduce the necessary muscular activity in targeted areas on the one hand, but might cause an actual increase in total body strain due to its weight or internal friction on the other hand. Simply measuring the reduction in muscular activity at the targeted muscle groups without any additional parameters might therefore lead to false reports on the overall usefulness of the exoskeleton.

What is needed is a global, overall indicator of how much stress is put onto the test subject by the work task and by how much the exoskeleton might help in reducing this stress despite causing an additional load by its weight which has to be carried additionally by its user [Pan16].

The easiest parameter to check for this is the heart rate. The heart rate (and its difference between conditions) is a formidable global parameter to find out how the work load (and additional load by the exoskeleton) will affect the user's cardiovascular system. A heart rate monitor is cheap, easy to apply and its interpretation rather uncomplicated even for non-physiologists or biomechanics experts. Another way to look at individual strain is by using specific questionnaires (see the section below).

4.5 Spiroergometry

Spiroergometry is a method of analyzing physiological parameters such as respiratory gas exchange, heart rate, etc. As such, it is not a biomechanical analysis per se, but since it is often used in combination with biomechanical methods such as 3D-kinemetry and EMG-analysis we will present its general approach here. Spiroergometry can be used to identify the individual cardiopulmonary and metabolic response to a given task. In medical assessments, spiroergometry is often combined with electrocardiography (ECG) but in the assessment of physical support systems such as exoskeletons it is mostly used to determine the amount of physiological effort during a certain task [Gal17].

In a nutshell, what spiroergometry does is the detection of O_2 and CO_2 during respiration. As the metabolism will change during exercise or work task, so will the relation between O_2 and CO_2. Additional to other outcome variables such as the amount of respiratory gas exchange during every respiratory cycle and heart rate (via a heart rate sensor) these variables are a product of the individuals metabolism and will thereby give insight into the cardiopulmonary impact a given task will have on the individual. As with EMG-analysis, these parameters are highly individual and depend on, e.g., fitness level, age, and gender. Inter-individual comparisons are only feasible when intra-individual changes between a baseline and test condition are compared—and not by any comparison of the outcome parameters themselves or when the values are normalized against, e.g., the body weight.

By comparing the outcome parameters between an unsupported and a supported task, it is possible to quantify the difference of global effort in between conditions. If a user's cardiopulmonary system and metabolism will benefit of a physical support system, the spiroergometry should provide the answer, as it will give a direct insight into the cardiopulmonary response of the user. However, as good as such a global outcome parameter of individual response may sound like, it is still not sufficient enough to be used solely. Even if the use of an exoskeleton for a given task will show reductions in cardiovascular strain in comparison to a performance without such an exoskeletal device, there are still aspects of possible kinematic or kinetic changes from one condition to the other to be considered. In the long run, alterations to body movements due to the exoskeleton might cause shear forces at the joints or postural deformities. This is why kinematic analysis plays such a vital role in the analysis of physical support systems.

4.6 Questionnaires

Even though questionnaires are no biomechanical analysis tools, it is most common to combine biomechanical analysis with a questionnaire in order to gain insight into user perceptions such as comfort, pain or other aspects of human-machine interaction.

Next to this kind of questionnaires, there are also a few standardized questionnaires which will help researchers to control for confounding factors in their test pool or complement biomechanical and physiological variables. Borg's RPE scale [Bor98], is for example one of the most used questionnaires in biomechanics to control for a homogenous strain on the test subjects induced by the task. The Ratings of Perceived Exertion (RPE) focusses on individual effort or strain while undertaking a certain task. The RPE-scale is a "relative" scale, asking about the "perceived" exertion caused by a certain activity and ranking it from a minimal to a maximal intensity [Bor06]. If, e.g., the RPE scales inside a test population or between groups will differ significantly, a comparison of physiological or biomechanical data must take into account, that the task at hand is most likely not of the same quality for all of the test subjects, meaning that even if the task itself is the same for everybody, individual differences like gender, height, strength, and experience will cause individual strain profiles [Ham09].

Another standardized questionnaire, often used to address homogeneity of a pool of test subjects is the Short Form Survey [Bul98]. This questionnaire measures eight different concepts regarding the individual health status like general health, bodily pain, vitality, etc. Because of its wide focus on a range of health-related aspects, it is mainly deployed as a tool to check for overall comparability in a test pool, filtering out subjects who may deviate too much as to be included in the test pool or to differentiate different groups inside a test pool. Clearly, those broader questionnaires are not meant to differentiate when it comes to specific questions such as if the test subjects are free of pain at specific body parts. The latter is critical, when for example an exoskeleton for the upper limbs is to be tested and the researchers have to make sure that the test subjects do not suffer from any kind of shoulder ailment. In these cases, researchers will rely on specific clinical assessments and questionnaires. However, in biomechanical analysis, the use of one or two specific questionnaires is most likely necessary to control for confounding factors.

5 Discussion

The evaluation of how physical support systems will interact with the human user is a major contributor in designing and improving these systems. Most of all, biomechanical analysis has to answer questions of how much the use of support systems will benefit their users. But one often neglected question is how much these systems will act as an additional load or alteration of human movements, resulting in unwanted harmful interactions [Hil17]. Each of the different methods of biomechanical analysis mentioned in this chapter can provide answers for certain aspects of the interactive support relation of human and machine while wearing an exoskeleton. Therefore, a set of different methods have to be combined to come up with an appropriate analysis of a users' interaction with a physical support system.

A combination of motion capturing methods, EMG-analysis and dynamometry, e.g., is capable of delivering answers about the amount of reduction of muscular activity by wearing an exoskeleton accompanied by an analysis of how much additional load has to be carried by the user and how movements will differ compared to the unsupported task at hand [Mal15]. Clearly, regarding the technical and time effort, these analysis combinations are mostly done in a laboratory. While it should be considered good scientific practice to match work tasks in the lab as closely as possible to their real-life counterparts, the former will always be influenced by an abundance of additional environmental and social factors. This is why in addition to biomechanical analysis of support systems in the laboratory, the real-life applications of these systems have to be analyzed as well to compare and validate the findings made in the lab. As [McG17] pointed out, even as tests in a laboratory will state that the use of an exoskeleton will be biomechanically favorable as, e.g., joint moments or loads on the human body will be reduced [Jac17], it is by no means certain that such a device is clinically indicated or will thereby automatically reduce musculoskeletal issues at the work place. To answer these questions prolonged field tests are needed—including biomechanical and physiological analysis at the workplace in conjunction with expert screening and user feedback.

Acknowledgements This research is part of the project "smartASSIST—Smart, AdjuStable, Soft and Intelligent Support Technologies" funded by the German Federal Ministry of Education and Research (BMBF, funding no. 16SV7114) and supervised by VDI/VDE Innovation + Technik GmbH.

References

[Alk01]   Alkjaer, T., Simonsen, E. B., & Dyhre-Poulsen, P. (2001). Comparison of inverse dynamics calculated by two- and three-dimensional models during walking. *Gait & Posture, 13*(2), 73–77.

[Bec17]   Beckerle, P., Salvietti, G., Unal, R., Prattichizzo, D., Rossi, S., & Castellini, C. (2017). A human-robot interaction perspective on assistive and rehabilitation robotics. *Frontiers in neurorobotics, 11*, 24.

[Bor98]   Borg, G. (1998). *Borg's perceived exertion and pain scales.* Chicago: Human kinetics.

[Bor06]   Borg, E., & Kaijser, L. (2006). A comparison between three rating scales for perceived exertion and two different work tests. *Scandinavian Journal of Medicine and Science in Sports, 16*(1), 57–69.

[Bul98]   Bullinger, M., & Kirchberger, I. (1998). *Fragebogen zum Allgemeinen Gesundheitszustand SF12.* Göttingen: Hogrefe.

[Bux12]   Buxi, D., Kim, S., van Helleputte, N., Altini, M., Wijsman, J., & Yazicioglu, R. F. (2012). Correlation between electrode-tissue impedance and motion artifact in biopotential recordings. *IEEE Sensors Journal, 12*(12), 3373–3383.

[Cem14] Cempini, M., Marzegan, A., Rabuffetti, M., Cortese, M., Vitiello, N., & Ferrarin, M. (2014). Analysis of relative displacement between the HX wearable robotic exoskeleton and the user's hand. *Journal of Neuroengineering and Rehabilitation, 11*(1), 147.

[Cho13] Chowdhury, R. H., Reaz, M. B., Ali, M. A. B. M., Bakar, A. A., Chellappan, K., & Chang, T. G. (2013). Surface electromyography signal processing and classification techniques. *Sensors, 13*(9), 12431–12466.

[Du16] Du, F., Chen, J., & Wang, X. (2016). Human motion measurement and mechanism analysis during exoskeleton design. In *2016 23rd International Conference on Mechatronics and Machine Vision in Practice* (M2VIP), IEEE, (pp. 1–5).

[Fau09] Faul, F., Erdfelder, E., Buchner, A., & Lang, A. G. (2009). Statistical power analyses using G*Power 3.1: Tests for correlation and regression analyses. *Behavior Research Methods, 41*, 1149–1160.

[Fra14] Franz, J. R., & Kram, R. (2014). Advanced age and the mechanics of uphill walking. A joint-level, inverse dynamic analysis. *Gait & Posture, 39*(1), 135–140.

[Gal17] Galle, S., Malcolm, P., Collins, S. H., & de Clercq, D. (2017). Reducing the metabolic cost of walking with an ankle exoskeleton. Interaction between actuation timing and power. *Journal of Neuroengineering and Rehabilitation, 14*(1), 35.

[Hal12] Halaki, M., & Ginn, K. (2012). Normalization of EMG signals: To normalize or not to normalize and what to normalize to? In *Computational intelligence in electromyography analysis-a perspective on current applications and future challenges*. InTech.

[Ham11] Hamaoka, T., McCully, K., Niwayama, M., & Britton, B. C. (2011). The use of muscle near-infrared spectroscopy in sport, health and medical sciences: Recent developments. *Philosophical Transactions. Series A, Mathematical, Physical, and Engineering Sciences, 369*(1955), 4591–4604.

[Ham09] Hamberg-Van Reenen, H. H., van der Beek, A. J., Blatter, B. M., van Mechelen, W., & Bongers, P. M. (2009). Age-related differences in muscular capacity among workers. *International Archieves of Occupational and Environmental Health, 82*(9), 1115–112.

[Hil17] Hill, D., Holloway, C. S., Morgado-Ramirez, D. Z., Smitham, P., & Pappas, Y. (2017). What are user perspectives of exoskeleton technology? A literature review. *International Journal of Technology Assessment in Health Care, 33*(2), 160–167.

[Hwa15] Hwang, B., & Jeon, D. (2015). A method to accurately estimate the muscular torques of human wearing exoskeletons by torque sensors. *Sensors, 15*(4), 8337–8357.

[Jac17] Jackson, R. W., Dembia, C. L., Delp, S. L., & Collins, S. H. (2017). Muscle-tendon mechanics explain unexpected effects of exoskeleton assistance on metabolic rate during walking. *The Journal of Experimental Biology, 220*(Pt 11), 2082–2095.

[Jar12] Jarrasse, N., & Morel, G. (2012). Connecting a Human Limb to an Exoskeleton. *IEEE Transactions on Robotics, 28*(3), 697–709.

[Jon16] Jones, S., Chiesa, S. T., Chaturvedi, N., & Hughes, A. D. (2016). Recent developments in near-infrared spectroscopy (NIRS) for the assessment of local skeletal muscle microvascular function and capacity to utilise oxygen. *Artery research, 16*, 25–33.

[Loo16] de Looze, M. P., Bosch, T., Krause, F., Stadler, K. S., & O'Sullivan, L. W. (2016). Exoskeletons for industrial application and their potential effects on physical work load. *Ergonomics, 59*(5), 671–681.

[Mal15] Malcolm, P., Quesada, R. E., Caputo, J. M., & Collins, S. H. (2015). The influence of push-off timing in a robotic ankle-foot prosthesis on the energetics and mechanics of walking. *Journal of Neuroengineering and Rehabilitation, 12*(1), 21.

[McG17] McGibbon, C. A., Brandon, S. C. E., Brookshaw, M., & Sexton, A. (2017). Effects of an over-ground exoskeleton on external knee moments during stance phase of gait in healthy adults. *The Knee, 24*(5), 977–993.

[Ott18] Otten, B. M., Weidner, R., & Argubi-Wollesen, A. (2018). Evaluation of a novel active exoskeleton for tasks at or above head level. *IEEE Robotics and Automation Letters, 3*(3), 2408–2415.

[Pan16] Panizzolo, F. A., Galiana, I., Asbeck, A. T., Siviy, C., Schmidt, K., Holt, K. G., et al. (2016). A biologically-inspired multi-joint soft exosuit that can reduce the energy cost of loaded walking. *Journal of neuroengineering and rehabilitation, 13*(1), 43.

[Ras14]   Rashedi, E., Kim, S., Nussbaum, M. A., & Agnew, M. J. (2014). Ergonomic evaluation of a wearable assistive device for overhead work. *Ergonomics, 57*(12), 1864–1874.

[Rob17]   Robert-Lachaine, X., Mecheri, H., Larue, C., & Plamondon, A. (2017). Validation of inertial measurement units with an optoelectronic system for whole-body motion analysis. *Medical & Biological Engineering & Computing, 55*(4), 609–619.

[Ruk16]   Rukina, N. N., Kuznetsov, A. N., Borzikov, V. V., Komkova, O. V., & Belova, A. N. (2016). Surface electromyography: its role and potential in the development of exoskeleton. *Sovermennye Tehnologii V Medicine, 8*(2), 109–117.

[Sin09]   Singh, T., & Koh, M. (2009). Effects of backpack load position on spatiotemporal parameters and trunk forward lean. *Gait & Posture, 29*(1), 49–53.

[Str17]   Strube, E. M., Sumner, A., Kollock, R., Games, K. E., Lackamp, M. A., Mizutani, M., et al. (2017). The effect of military load carriage on postural sway, forward trunk lean, and pelvic girdle motion. *International Journal of Exercise Science, 1*(10), 25–36.

[Val18]   Valevicius, A. M., Jun, P. Y., Hebert, J. S., & Vette, A. H. (2018). Use of optical motion capture for the analysis of normative upper body kinematics during functional upper limb tasks. A systematic review. *Journal of Electromyography and Kinesiology: Official Journal of the International Society of Electrophysiological Kinesiology, 40*, 1–15.

Mass Survey for Demand Analysis

Alexander Mertens, Katharina Schäfer, Sabine Theis, Christina Bröhl,
Peter Rasche and Matthias Wille

Abstract In the ongoing digitalization of society new technical systems and technologies are increasingly penetrating people's everyday lives. In order to be able to analyze the resulting complex interactions and forms of networking, a participative approach is needed to identify the needs of these user groups. Empirical studies, e.g., mass studies, are important because it may be required that many stakeholders have to be questioned in a short period. In this article, various methodological approaches are presented using best practice examples to show the strengths and weaknesses of these methods.

1 Introduction

Technical systems and new technologies permeate our everyday lives more and more and are a fundamental part of our environment. The sustainable development of appropriate systems and products for the integration into the respective socio-technical systems requires a participatory approach. Through this practice both the needs and desires but also the fears and obstacles of potential future can be considered. This topic is of particular importance due to the increasing complexity of interaction and networking of people with technical components. It may be necessary to involve and question a large number of stakeholders in a short period to create user profiles, e.g., to generalize or validate results from empirical studies of smaller samples in the laboratory or field. In order to support an usage that is independent of a specific discipline and possible application-contexts different methodological approaches and systematics of this topic will be briefly explained below, according to their use in the human-centered development cycle DIN EN ISO 9241-210. Concrete best practice experiences for different implementation variants will also be presented, which consider the strengths and weaknesses of the respective methods.

A. Mertens (✉) · K. Schäfer · S. Theis · C. Bröhl · P. Rasche · M. Wille
Institute of Industrial Engineering and Ergonomics, RWTH Aachen University, Aachen, Germany
e-mail: a.mertens@iaw.rwth-aachen.de

© Springer Nature Switzerland AG 2018   63
A. Karafillidis and R. Weidner (eds.), *Developing Support Technologies*, Biosystems &
Biorobotics 23, https://doi.org/10.1007/978-3-030-01836-8_6

1.1 Theoretical Background

Since the emergence of empirical research, different paradigms have maintained the focus on social development from different perspectives. Together, they aim to collect a massive range of data in order to be able to make statements on their specific subject area [Vis14]. It must be carefully considered which design can best cover or answer a specific question, because different research designs can lead to different results [Zha17]. In this context, mass surveys represent a large part of empirical research. They are able to anticipate social trends in a broad group and make them measurable by means of statistical methods [Sch49]. Also, they are able to verify or falsify hypotheses and to generalize statements about the relevant population. Mass surveys are necessary for participatory requirements analysis, as a large number of different actors have to be questioned about new developments in a short period of time in order to identify possible needs for action and to develop target group specific design recommendations. The procedures described below can be used to ensure that the results obtained in the studies have a certain degree of generalizability and that other research groups with a similar background can take up these results.

1.2 Mass Surveys as Part of a Human-Centered Development Process

The participatory conception, implementation, and evaluation of new technical systems and technology-supported services is divided into four concrete steps and associated usability engineering activities in accordance with the "human-centered development process" of DIN EN ISO 9241-210: (1) Understanding and describing the context of use, (2) specifying usage requirements, (3) developing design solutions, and (4) designing solutions. By purposefully iterating these steps until the design solution meets the usage requirements, the design can be based on a comprehensive understanding of the users, work tasks, and the working environment and also take complex reciprocal socio-technical dependencies into account. By using methods of mass surveys an extremely effective and efficient realization can be achieved, especially for steps (1), (2), and (4). In order to decide which method can be used meaningfully, the time dimension has to be taken into account—whether it is a long-running process that provides for regular iterations and revisions of the technical components or whether it is a completed development process that is no longer adapted after finalization of the specification. Depending on this, methods from either the field of longitudinal studies or from the field of cross-sectional studies are preferable (see Fig. 1).

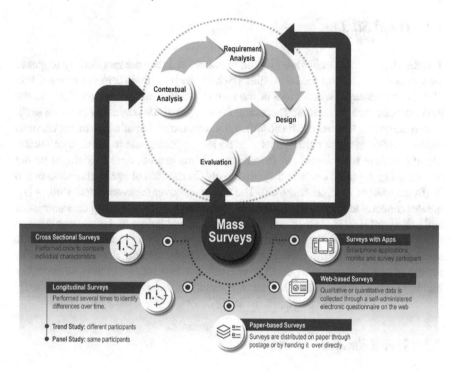

Fig. 1 Methods of mass surveys applied in the human-centered development process

In the following sections, concrete best practices for the use of mass surveys from application-oriented practice will be presented and discussed with regard to their strengths and weaknesses. The aim was to identify the requirements for the design of innovative technical systems and technology-based services with a "large" sample size in all examples.

2 Longitudinal Studies

Longitudinal studies are designed to investigate changes in social and individual processes over time. In contrast to cross-sectional studies, longitudinal studies are conducted at multiple points in time and the results are compared between the different times. Two different forms of longitudinal studies are distinguished: the panel study, where the same sample is used at each point in time, and the trend study, where different samples are used at different points in time.

2.1   Panel Studies

The data from panel studies are difficult to collect and they are more costly to gather than data from cross-sectional studies [Hsi07]. Beyond that there is always a loss of data, for example when some of the former participants are not available at the next point of data collection. Therefore, the sample at the beginning must be sufficiently large that the loss can be absorbed. However, a systematic bias might occur as characteristics of those participants who no longer participate in the data collection might correlate with aspects of the object of investigation, e.g., they might be not participating anymore because they are too old (in the case of age differences) or too ill (in the case of medical research). The advantage over cross-sectional studies is a greater capacity to display the complexity of human beings, including characteristics and behavior [Hsi07]. Also, panel data allow a direct "before and after" comparison on an individual basis when a within-subject design is given, since data exists for the same person at different points in time. In studies concerning decriminalization or the use of new therapy forms, for example, this is indispensable. Finally, such studies highlight dynamic relationships, that is, correlations can be found and interpreted.

2.2   Trend Studies

Contrary to panel studies, trend studies contain different samples, so many more participants are needed overall. The acquisition of participants is thus more time-consuming and costly. But as a benefit, every sample is tested only once, and no potential systematic loss exists like it does with panel studies. However, another kind of systematic bias within the sample might occur as in all voluntary surveys: participants with less interest in the topic of the survey are less likely to answer the voluntary questionnaire. With the results of a trend study it is possible to interpret variations in data based on the whole sample, which should represent the population. It is not possible, as it is in panel studies, to attribute variations directly to interindividual differences, as here a between-subject design is used. In this form of investigation, the data of different samples is collected at each round of implementation so that not individual changes but rather generic differences are identified. In order to achieve representative and generalizable results, however, the respective sample, the so-called cohort, is built up in a way to be as comparable as possible, for example with regard to age, gender, social, and cultural background. It can be stated in conclusion, that if the main research focus is on evolving trends of, for example, technology use over time rather than, for example, individual learning effects, a trend study is more applicable than a panel study. Panel studies are only important if the focus is on individual change over time.

The first challenge when setting up a trend study is to involve a large number of participants who are representative for the population. One good way is to contact professional address providers, who may even sort or restrict your sample to conditions important for the research, like a specific age group. For the long-term trend study of the "Tech4Age" project [Wil16, Mer17] the authors used the service of DHL and sent a 26-page paper-based questionnaire to 5000 people of 60 years or older equally distributed over the whole of Germany. Initial case numbers in the thousands are normal in trend studies, as only 10–20% of the questionnaires are returned. Sending out such huge numbers of questionnaires entails a lot of handwork for a small research group, which might be another reason to pass this over to a professional service provider. The response rate will depend on the topic, size, and appearance of the questionnaire. Therefore, it is recommended to keep the questionnaire as short as possible and frame it with a covering letter that explains the topic and the necessity of the research in short and comprehensible sentences. A postage-paid envelope should also be included to avoid costs for the participants. As an incentive it should be mentioned in the cover letter how important the opinion of the participant is for that field of investigation. Some surveys also offer a prize draw, giving those people who return the questionnaire the opportunity to win a small prize. Nevertheless, return rates over 15% are rare for these voluntary random requests, so ten times the number of surveys should be distributed than answers required. Overall, the bigger the sample is, the better and more trustworthy the results will be. Trend studies might contain both qualitative and quantitative questions. However, as big sample sizes are to be dealt with, the analysis of qualitative data would be enormously time-consuming. Therefore, it is better to concentrate on quantitative research and use open questions and qualitative research only if necessary. If the trend study is paper-based, it should be kept in mind that all incoming questionnaires have to be transferred to electronic media for further data analysis. It is thus preferable to handle this by means of an automatic read-in process during the scanning of the returned questionnaires. There are programs specialized in this approach (e.g., Remark), but the questionnaire has to follow a specific layout in terms of size, position, and answer categories for this to work properly. This has to be considered when designing the questionnaire.

3 Cross-Sectional Studies

In empirical research, one speaks of a cross-sectional study or cross-sectional design when an empirical investigation is performed once. In contrast, a longitudinal study is performed several times in succession. Cross-sectional studies compare the results of each participant and provide information about the prevalence of a behavior or attitude. Moreover, the data can provide a "snapshot" of individual characteristics at a specific time. The procedure of a cross-sectional study is generally quicker, easier, and cheaper compared with longitudinal studies. Furthermore, there is no data loss, due to the fact that one participant is only interviewed once [Sed14]. But simultaneously this limits the studies because they do not permit indications of data differences over time [Lev06]. This makes it hard to find causal correlations. The obtained values form the dependent variables.

4 Best-Practices for Methods of Mass Surveys

4.1 Web-Based Surveys

A web-based survey can be used to conduct qualitative as well as quantitative research [Eys02]. Data is collected through a self-administered electronic set of questions on the web. Web-based surveys are used for data collection and should be incorporated when the study design as well as the questionnaire are fully defined and developed.

Depending on the electronic tool used to conduct the survey, its physical appearance and presented information can usually be adapted easily. Web-based surveys additionally offer the potential to create a dynamic questionnaire, that is, questions are hidden or revealed depending on the former answers of participants. This prevents the participants from getting bored or frustrated with questions they can not or do not want to answer. Also, further specific questions can be asked, making the collected data much more informative. Regarding data quality, web-based surveys have further advantages. The used electronic tool can monitor data input and give feedback on whether a question is missed or data input is wrong. These are some reasons why web-based surveys have been claimed to reduce respondent error and increase the completeness of responses.

But these functions also come with disadvantages. Participants might be "forced" to answer certain questions to proceed and therefore might not choose the most suitable answer, but simply click randomly in order to proceed. This shows that the design of web-based surveys needs to be tested extensively as the electronic questionnaire is the sole form of communication between investigator and participant. To detect incorrect inputs, the software needs to know which answers are valid. For quantitative studies with closed-ended questions this is quite easy to determine. But in the case of qualitative research with a lot of open-ended questions it becomes a lot more difficult to determine the right answer and therefore use this error detection function. Our experience has shown that web-based surveys should be used for quantitative studies with few open-ended questions, the answers to which can be determined quite well in terms of type as well as possible length [Ras17]. The use of a web-based survey to conduct qualitative research with open-ended questions should yet be considered, as this type of survey avoids the transcription of handwritten answers for analysis.

Web-based surveys are suitable for reaching a large group of participants with a small team of investigators. Therefore, and because of the instant recording of data, web-based surveys are frequently described as potentially time-, effort-, and cost-saving.

The recruitment of participants for web-based surveys can be done in different ways. There are several examples with quite good results recruiting via web, e-mail, or social media [Ras17, The17, Bro16]. Web-based surveys are less intrusive as participants do not need to answer the survey immediately but can rather choose a time and place to answer the survey free from stress. Research has shown that recruitment via web, e-mail, or social media is not representative regarding a whole population but is suitable for reaching individuals with particular characteristics or interests, e.g., potential users of a certain game, experts in robotics, or users of electronic health records [Bes04, Top16]. Users of social media platforms like Facebook are adequate in terms of representative population characteristics [Rif15]. Limited access to the internet, security issues, and technical problems might also indicate differences among participants.

All in all, web-based surveys are a suitable and cost-efficient way to conduct quantitative research among individuals with specific characteristics in a short period of time.

4.2 Paper-Based Surveys

Paper-based surveys, as the name already implies, are surveys that are distributed on paper via mail or by handing the surveys directly to participants. Because of the heavy use of online surveys in recent years, paper-based surveys are sometimes seen as outdated, although there are still advantages to paper-based surveys in comparison to online-based designs.

One advantage is the fact that not every age group addressed in a study uses computers and online services on a regular basis. People aged between 50 and 75 years show significantly less initiative to complete online surveys and prefer the use of the paper-based form [Bec09]. In order to account for visual impairments of older people, the font of the paper-based surveys can be scaled to an appropriate size, without affecting display problems by deforming or distorting the typeface on the screen. An example of an analysis which often needs a sample that spans multiple different age groups and where a paper-based survey would be appropriate is the Kano analysis. The Kano analysis is based on the Kano model, a model which is used to examine product features with regard to their impact in early stages of the development of innovative products and aims at effectively integrating different target groups into the developmental process [Bra16] for example used the Kano model to analyze different customer requirements with regard to a technical mobility aid. To categorize product features, a positively formulated question and a negatively formulated question are asked straight after each other. After several product features are assessed in this way the model is built and design recommendations can be deduced [Kan84].

The use of the paper-based form of survey is also appropriate for age-independent samples. Some surveys require the participation of a specific sample, e.g., experts of a specific field, which might be hard to find. In this case, it might be easier to give the survey in person to a subject, e.g., at a conference or a meeting, than sending the survey online via e-mail, as people usually have a pen to hand and are willing to fill out a survey on the spot at an event instead of answering online questions after an event. An example of a study based on expert ratings was published by [Bro13]. The researchers studied services with regard to medical care in the future and administered a Delphi study. A Delphi study is a systematic, multi-level analysis of experts' ratings. A panel of experts convenes and gives feedback on selected topics by means of a survey in two or more rounds. After each round an anonymized summary of the experts' ratings from the previous round as well as the reasons they provided for their ratings is created and given to the sample in the next round. It is believed that during this process the range of answers will decrease and the group will converge towards the "correct" answer. The Delphi study is an example where a paper-based survey would be highly recommended.

Although there are a lot of advantages of paper-based surveys, there are some disadvantages as well. Paper-based surveys deploy finite resources (including paper) and are therefore not environmentally friendly in large quantities. Moreover, the high cost for material, printing, and mailing are not to be underestimated and should be considered especially for large cohorts. The processing of the data sets also requires a number of additional resources, since the questionnaires must generally be digitized before the concluding evaluation. Finally, and foremost, paper-based surveys are not dynamic, or rather they can only be dynamic to a certain extent. That means that it is more complicated to form groups of people who answer questions based on their answers to previous questions, something which is easily achieved in web-based surveys through the integration of dynamic loops.

4.3 Large-Scale Surveys as Part of a Smartphone Application

Study designs to investigate user behavior and the effect of apps on users typically require direct contact between participants and examiner as part of participant recruitment and briefing, as part of the instruction of experimental hardware as well as during surveys. However, the direct contact can cause effects of social desirability on the participants' behavior and opinion, especially since health-related behavior is socially relevant. During investigations of usage duration and drop-out rates, further biases can result from specifying an end time of the experiment, by the users "persisting" despite the fact that their motivation has declined. Another problem is the Hawthorne effect. This effect is defined as a change in behavior resulting from the participant's consciousness about his participation in a study. The salience of the study situation is further increased by personal contact, the use of study-specific devices, the knowledge of an end date, and additional interviews.

One solution addressing the described drawbacks are surveys and behavior tracking by smartphone applications and persuasive self-monitoring systems. Since the recording of data is already an integral part of the monitoring system, these data can be evaluated in the context of a certain research question. The widespread use of smartphones makes it possible to let participants use their usual phone instead of unusual, study-specific devices. The core of the proposed method is denoted by providing a smartphone application in common app stores. As part of the installation process, the user is informed about their participation in the surveys and agrees to this and to the collection of their data. For comparative studies, one of several app versions may be randomly assigned during installation. Relevant data can be recorded as far as possible during use. Usage duration should be taken into account here.

The described process of recruitment and the exclusion of direct contact establish a distance between participants and examiner and thus ensure anonymity for the participants. This is important in order to avoid a feeling of being observed or evaluated, which could encourage the participants to act in what they consider to be a socially desirable way. The possibility for participants to use their own smartphones, the availability of established distribution paths (app store) as well as the exclusion of purely study-related activities and a temporal limitation increase the naturalness of the usage situation. Additionally, making the application available in an app store answers the question of whether participants would use the application in real situations. The composition of the user group with regard to the recorded data can also be determined without sample bias, although limited information might be available about the users' demographic variables in cases where not explicitly queried.

Experiences indicate that the Hawthorne effect and socially desirable behavior can be largely avoided with the chosen study design. However, while studies with a classical design report low dropout rates, smartphone-based surveys and data monitoring studies commonly face dropout rates of around 75% already after one week. This is comparable to results from media usage surveys, which show that on average 80% of all installed apps are not used. Another limitation of app-based investigations is a sample of participants who are already interested in the objective or who have higher affinity to digital technologies. This method is therefore only recommendable if the characteristics of users who download the app can be expected to match the characteristics of the target population. Furthermore, the final number of participants is difficult to predict, which is why sufficient time for advertising and PR of the app needs to be ensured.

In order to have valid experimental data, both active and passive usage should be recorded. In this case, high exclusion rates are to be expected to influence the quality of experimental data. Finally, the development effort is not to be underestimated. The smartphone application must be a market-ready product, with users expecting technical support as well as constant bug fixes and functionality extensions. It is important to consider that this approach is not suitable for investigations that require additional information to the data collected by the monitoring system. It is, however, suitable to investigate application usage and to compare different application components on use behavior tracked by the monitoring functions.

5 Conclusion

The aim of this chapter is to demonstrate different best practice examples in the field of mass surveys. A brief theoretical introduction to the subject of mass surveys and human-centered development designs was given before various forms of research in the field of mass surveys were presented. In a first step, longitudinal studies—trend and panel surveys—were discussed. In a second step, cross-sectional studies as an alternative approach were presented. The focus of the latter chapter was on web-based and paper-based surveys as well as surveys with smartphone applications to realizes large-scale surveys.

Regardless of the methodology used, it must be borne in mind that the data collected may contain personal information which must be treated with due care. In addition to the principles of data protection, such as data economy, earmarking, and transparency, strong anonymization procedures or, in the case of investigations in within-subject design, consistent pseudonymization can be used to ensure that the data can no longer be directly assigned to any person. In addition to the consideration of corresponding rules, which are a partial aspect of Responsible Innovation and Research (RRI), the ethical, legal, and social implications of research should be considered prospectively and all involved stakeholders should be taken into account in the sense of a continuous participation.

The overall aim was to link theoretical background with information and practical experience in the interdisciplinary research project Tech4Age (www.tech4age.de) and thus demonstrate the advantages and disadvantages of these methods.

Acknowledgements This publication is part of the research project "TECH4AGE," financed by the Federal Ministry of Education and Research (BMBF, under Grant No. 16SV7111) and promoted by VDI/VDE Innovation + Technik GmbH.

References

[Bec09] Bech, M., & Kristensen, M. (2009). Differential response rates in postal and Web-based surveys in older respondents. *Survey Research Methods, 3*(1), 1–6.

[Bes04] Best, S. J., & Krueger, B. J. (2004). *Internet data collection, Quantitative applications in the social sciences* (141st ed.). Thousand Oaks, California: Sage Publications.

[Bro13] Bröhl, C., Mertens, A., Brandl, C., Mayer, M., & Schlick, C. (2013). Integration technischer Assistenzsysteme in die personenbezogene Dienstleistungserbringung—Ergebnisse einer Delphi-Studie. In: Lebensqualität im Wandel von Demografie und Technik, Berlin: VDE-Verlag, (pp. 234–238), 2013.

[Bro16] Bröhl, C., Nelles, J., Brandl, C., Mertens, A., & Schlick, C. M. (2016). TAM reloaded: A technology acceptance model for human-robot cooperation in production systems. In C. Stephanidis (Ed.), *HCI 2016* (Vol. 617, pp. 97–103). CCIS Cham: Springer.

[Eys02] Eysenbach, G., & Wyatt, J. (2002). Using the internet for surveys and health research. *Journal of Medical Internet Research, 4*(2).

[Hsi07] Hsiao, C. (2007). Panel data analysis—advantages and challenges. *TEST, 16*(1), 1–22.

[Kan84] Kano, N., Seraku, N., Takahashi, F., & Tsuji, F. (1984). Attractive quality and must-be quality. *Journal of the Japanese Society for Quality Control, 14*(2), 147–156.

[Lev06] Levin, K. A. (2006). Study design III: Cross-sectional studies. *Evidence-based Dentistry, 7*(1), 24.

[Mer17] Mertens, A., Rasche, P., Theis, S., Bröhl, C., & Wille, M. (2017). Use of information and communication technology in healthcare context by older adults in Germany: Initial results of the Tech4Age long-term study. *i-com, 16*(2), 165–180.

[Ras17] Eysenbach, G., & Wyatt, J. (2002). Using the Internet for surveys and health research. *Journal of Medical Internet Research, 4*(2).

[Rif15]   Rife, S. C., Cate, K. L., Kosinski, M., & Stillwell, D. (2016). Participant recruitment and data collection through Facebook: The role of personality factors. *International Journal of Social Research Methodology, 19*(1), 69–83.

[Sch49]   Schneider, H. J. (1949). Voraussage durch Massenbefragung. *Wirtschaftsdienst, 29*(4), 22–26.

[Sed14]   Sedgwick, P. (2014). Cross sectional studies: Advantages and disadvantages. *BMJ: British Medical Journal, 348.*

[The17]   Theis, S., Rasche, P., Bröhl, C., Wille, M., & Mertens, A. (2017). User-driven semantic classification for the analysis of abstract health and visualization tasks. In V. G. Duffy (Ed.), *DHM 2017* (Vol. 10287, pp. 297–305). LNCS Cham: Springer.

[Top16]   Topolovec-Vranic, J., & Natarajan, K. (2016). The use of social media in recruitment for medical research studies: A scoping review. *Journal of Medical Internet Research: JMIR, 18*(11), e286.

[Vis14]   Visser, P. S., Krosnick, J. A., & Lavrakas, P. J. (2014). Survey research. In H. Reis & C. M. Judd (Eds.), *Handbook of research methods in social and personality psychology* (2nd ed, pp. 223–252). New York, NY: Cambridge University Press.

[Wil16]   Wille, M., Theis, S., Rasche, P., Bröhl, C., Schlick, C., & Mertens, A. (2016). Best practices for designing electronic healthcare devices and services for the elderly. *i-com, 15*(1), 67–78.

[Zha17]   Zhang, X., Kuchinke, L., Woud, M. L., Velten, J., & Margraf, J. (2017). Survey method matters: Online/offline questionnaires and face-to-face or telephone interviews differ. *Computers in Human Behavior, 71,* 172–180.

The Burden of Assistance.
A Post-phenomenological Perspective
on Technically Assisted World Relations

Bruno Gransche

Abstract Technology is not just a useful tool to achieve certain goals, it is also a medium to relate to the world. The aim of this article is to examine the specific transformational power of comprehensive technical assistance systems in terms of our world relation. It applies Don Ihde's early concepts of embodiment, hermeneutic, and background relations to today's highly automated assistance systems and examines the interplay of human actors and the technosphere in a post-phenomenological perspective. In a lifeworld context assistance is automatically seen as a relief. Hence this article raises the question of what the burden of assistance might be.

Technical support systems are constantly growing in terms of performance and autonomy, and so is the number of tasks we can delegate to technology. To put it briefly, if machines can achieve more, we can in turn use these machines to achieve more, too. With regard to means-to-an-end relations, the more useful technology becomes as a realm of means, the more the realm of ends grows. Since useful means and achievable ends are co-constitutive concepts, this means that an increasing number of phenomena are subject to a machine-driven modal shift. What was once thought to be practically impossible now becomes practically possible in the course of technological progress. What was once expressed as a wish now becomes achievable, i.e., a possible purpose of actions. Technically assisted actions have different purposes than unsupported actions. Support services were developed to expand this dimension in the first place. Today, when human beings take action, they always interact with other co-actors and a network of assisting structures. Humans no longer exclusively cooperate with other human actors, but also with technical agents within a *technosphere*. In the case of intelligent highly automated assistance systems, acting in and with technology has considerably more consequences than simply expanding the realm of purposes. Technology is not just a useful tool to achieve certain goals, it is also a medium to relate to the world.

B. Gransche (✉)
Institute of Advanced Studies FoKoS, University of Siegen, Weidenauerstr. 167, 57076 Siegen, Germany
e-mail: bruno.gransche@uni-siegen.de

© Springer Nature Switzerland AG 2018
A. Karafillidis and R. Weidner (eds.), *Developing Support Technologies*, Biosystems & Biorobotics 23, https://doi.org/10.1007/978-3-030-01836-8_7

The aim of this article is to examine the specific transformational power of comprehensive technical assistance systems in terms of our world relation. The foundation for this is Don Ihde's early approach to philosophy of technology, which considers human-machine relations to be of hermeneutic nature and looks at amplifying and reducing effects. First, this article summarizes the difference between embodiment and hermeneutic relations, as well as background relations according to Ihde. Then, the focus lies on the special relation between highly automated assistance systems and the aspect of background relations; thus, technical assistants appear in the modality of a "deistic god" [Ihd79, p. 14] with a presumption toward totality. Machine-mediated experiences of the world, as well as experiences with machines as a world, feature an interplay between amplification and reduction. Finally, the objective is to examine this interplay in a post-phenomenological approach. The hypothesis of this article is that when it comes to using technology, the focus usually only lies on one type of effect—either extension or reduction—whereas the other is taken for granted and therefore disregarded. This natural process of overlooking the other effect is to be put back into focus here. Assistance literally means the action of helping someone by sharing work. However, if this action of helping out has complex effects on the world relation of the assisted, we have to ask the following question while considering its full complexity: What burdens are actors relieved of and what burdens are imposed on them? If, in a lifeworld context, assistance is automatically seen as a relief, we cannot avoid the question of what the burden of assistance is.

1 Embodiment Relations—
Hermeneutic Relations—Background Relations

Phenomenology is a philosophical study in the early 20th century that puts the spotlight on the *phenomenon*, i.e., the appearance, the givenness of an object to the active consciousness. The question was no longer what an item objectively is, but rather how it is perceived by one individual's consciousness. In *Technics and Praxis* [Ihd79], Don Ihde lays the foundation for a phenomenologically substantiated philosophy of technology, that has been developed further into post-phenomenology since then and constitutes a highly dynamic concept in philosophy today. This focus on the givenness for someone's consciousness mainly addresses the relation between humans and the world. In current times, this relation is mainly machine-mediated, making it a human-machine-world relation. For example, when a dentist scans the surface of a tooth using a probe, he mainly perceives the surface area through the use of technology. This mediation coincides with a type of transformation. Some impressions—e.g., small bumps on the tooth's surface—are amplified and others—e.g., temperature—are reduced. This relation signifies an "experience *through* machines" [Ihd79, p. 56]. It can be seen as a type of embodiment relation, since the technical instrument becomes partially or completely transparent in this kind of perception. In the given example, the dentist appears to directly feel with the probe and ignores the instrument that mediates between the tooth and the hand in the process. Here, we always face a twofold transformation of perception: a sensory extension and reduction at the same time. The instrument is used as an extension while the aspect of reduction can be set aside depending on the respective interest. Focussing on extension, disregarding reduction, and integrating an instrument to become part of one's own body: all of these are steps towards machines being taken for granted. Husserl, founder of philosophical phenomenology, refers to the universe of what is taken for granted without question as *Lebenswelt* (lifeworld). Phenomenology aims to take apart what is taken for granted and systematically uncover the full complexity of the givenness. Ihde refers to this human-machine relation of experience through machines as *embodiment relation* and contrasts naked perception with machine-mediated experience.

Another type of human-machine relation is when humans do not experience the world by means of technology, but rather see the world as represented by technology. For example, this refers to any representation of physical sensory data—such as temperature, air pressure, traffic density, etc.—on a screen. This is the case in control centres of large-scale plants or on a computer screen. If we look at a symbolic output of sensory data on a screen, our primary relation is not with the recorded reality but with the technical version of that reality. Ihde refers to this type of relation, in which a person experiences the technical representation of a world, as *hermeneutic relation*. This kind of relation is hermeneutic because the individual has to *interpret* the system's output and the visualization of the information. For example, a green or red control light indicates an underlying condition, such as the temperature being at the right level or too high. In the case of hermeneutic relations machine-mediated perception correlates with a type of transformation as well: depending on the system, the person perceives different details of the world. This means that a control light in a control center can provide information on the temperature of each individual sensor, but not on the temperature distribution between the sensors. Machine-mediated experience is always more and at the same time less than naked experience. To determine the degree to which technical transparency or opaqueness, as well as amplification or reduction are artefacts for machine mediation, one has to compare the machine-mediated experience and the naked perception of the same section of the world. In the case of an unlikely temperature, the observer could avoid machine mediation by entering the room containing the sensor in question. Potential for transparency and transformation can be found in the difference between naked perception and machine mediation. In most cases, due to an increase in human-machine relations, avoiding machine mediation by means of naked perception is not possible. For example, representations created by scanning electron microscopes or space telescopes display sections of the world that have never been and are never going to be accessible through naked perception. Here, humans have to compare different machine mediations to perceive information on transformational artefacts or characteristics of the world. This process of extrapolating and interpreting the world and machines leads to humans existing in a hermeneutic relation to both.

A third type of human-machine relation plays a central role here: due to technology being increasingly autonomous, interconnected, and capable of learning, humans are moving from an instrumental relation, in which they embody technology, over an operational relation, in which their perception is primarily focused on technical systems and their interfaces, to a type of relation, in which they only momentarily regulate an otherwise mostly automated *technosphere*.

1. Embodiment relations	(Human-machine) → world
2. Hermeneutic relations	Human → (machine-world) [Ihd79 p. 13]
3. Background relations	Human → $\left(\frac{\text{machine}}{\text{world}} \right)$

A wide variety of technical systems form a variety of *background* relations. Human actors only encounter these *background* relations when, e.g., setting goals, aborting a mission, or in case of malfunction. When it comes to transformation and transparency, the *technosphere* of interconnected and highly automated systems displays specific tendencies: as the term *background* relations indicates, relations between systems do not take place in the foreground. Most of the time, these relations are *transparent*, in the sense that they are invisible to humans or normally not perceived like window glass. Furthermore, apart from a few exceptional cases, transformations of the *technosphere* relation can no longer be translated into naked perception. To human actors, the *technosphere* appears in the modality of a "deistic god".

> In these cases I have had a momentary relation with a machine, but in the modality of a deistic god. I have merely adjusted or started in operation the machinery which, once underway, does its own work. I neither relate through these machines, nor explicitly, except momentarily, to them. Yet at the same time I live in their midst, often not noticing their surrounding presence. [Ihd79, p. 14]

This means that the *technosphere* is mostly unavailable to humans and does not have any intentions towards them. Inexplicably defiant printers or elevators, for example, do not hold any scrutinising or chastising intentions of a personal, theistic god—although these traits may be assigned as part of an anthropo- or theomorphic projection every day. Instead, these machines could be regarded as an almost ubiquitous creation that, once it has been launched, operates autonomously.

Rather than making naked experiences in direct human-world or human-nature relations, we spend the majority of our lives in the three human-machine-world relations discussed here. Today, we can hardly imagine living in a world or even spending a day without technology. Since our culture is shaped by technology, machine mediation is part of our self-awareness and self-expression. Technical systems have become our familiar counterparts as quasi-others, and they surround us with their presence from which we rarely escape. As a technological texture to the world and in the modality of a deistic god, the *technosphere* has a tendency towards totality. To live means to live with technology and being-in-the-world means being in technology with technology [Ihd79, p. 15].

2 Specific Relations Between the Assisted and Assistance Systems

If we apply this existential structure of general human-machine-world relations to the example of advanced assistance systems, we can examine certain dynamics and consequences that become relevant when people are using assistance systems.

Considering this structure, we can ask the following questions:

- What type of relation does each assistance situation primarily belong to?
- What are the present aspects of embodiment?
- How transparent or opaque is the machine mediation?
- What are the present transformation, amplification, and reduction effects?
- What are the present *background* relations?

These questions can help analyze existing assistance systems and can act as a heuristic approach when making decisions on the design and development of new systems. For this, we first have to ask to what extent an application *should* be transparent (in the above-mentioned sense of non-perceivability) to the user and what the intention behind it is.

Generally speaking, we can assume that, unlike tools or machines that are only automated to a small extent, assistance systems tend to lean less towards embodiment relations and more towards hermeneutic relations with strong *background* relations. To illustrate this, let us compare two examples of assistance systems: First, exoskeletons, that are the focal point of systems in the project smartASSIST[1] and that provide physical strength in assembly work. Second, highly automated assistance systems used in everyday life, such as Google Home or Amazon Alexa.

In the case of exoskeletons used in overhead assembly work, these systems carry the majority of the weight of the workers' arms or of the workpieces. Unlike the embodied probe working on a tooth, there is no reference to the world in the sense that the naked perception of one's own hands would be transformed by instruments. Furthermore, the hands of the worker are the ones guiding the tools and the workpiece. The objects' temperature, consistency, or structure are not transformed in this process. However, one characteristic is reduced: the weight of said objects. The worker's hands still have to grip with an adequate amount of manual labour, they provide information on the differences in weight, but the lifting capacity is supported by the system. This transforms how this characteristic is perceived, weight is reduced and endurance as well as performance in overhead construction are improved. Depending on the system's design, the strength dynamics of the assistance, and the hardware's ergonomics, the embodied perception of the exoskeleton can vary. In this case, the system and the design aim at providing an authentic human-thing or human-world relation while reducing physical effort at the same time. In doing so, the assistance system should be as transparent as possible, the transformation of perception should be limited to selected aspects and the degree of embodiment should be rather high.

---

[1] http://www.humanhybridrobot.info/smart-assist/, last checked 02.01.2018.

This also implies that the exoskeleton mostly should not move into the user's focus as a technical counterpart. This affects the hermeneutic relation between the user and the system: Strength assistance is supposed to entirely depend on the user's control, meaning that the user does not have to interpret the system's feedback in terms of world mediation. This is different with sensitive prosthetics or teleoperated robots, where the signals of the system have to provide information on an item's condition. In terms of the system that is providing individual support with one system each, *background* relations are of lesser importance. Even in the case of a system capable of learning, other users' movement patterns would only provide little information on how to offer adequate individual support. A connection to another network would only make sense when recording and analyzing data. To summarize, in the case of exoskeletons, human-machine-world relations can be found in the form of embodiment relations, preferably featuring isolated transformation and high transparency.

Examining our second example, everyday assistance systems such as Google Home, we are presented with a different picture. Both the assistance system and the interface devices are not in physical contact with the user. The Google Home interface mainly consists of microphones. The hands-free voice control is the assistant's main unique selling point. Due to this physical distance, there is no embodiment relation. The dialogue-based voice control creates a technical counterpart and therefore leads to a hermeneutic relation. An interpretation of the world reflected by the system and the system's condition is mainly based on spoken interaction, even when visual aspects such as the screen can sometimes factor in as well. Assistance systems such as Google Home, Alexa by Amazon, Windows Cortana or Siri by Apple give the impression that we are interacting with a quasi-human counterpart. However, what really makes these systems so efficient is a wide variety of *background* relations that hide behind a simulated interlocutor. Therefore, the skill set of those systems—e.g., Alexa's Skills—mainly depends on their being interconnected with each other as well as a large number of sensors and actuators. Thus, different requests heavily depend on the sensors. A system can only answer the question "Is it cold outside?" if it is connected to the respective temperature sensors. Similarly, a system can only answer the question "How long will it take me to get to work today?" if it has access to traffic information—via traffic sensors such as traffic cameras—and to information on the location of the office and the current location of the user. In the case of assistance systems, the user has to interpret the system's voice output and its connection and relevance to the world. Here, the perception of the world is transformed on many different levels: from the design, settings, placement, density, precision, and interconnectedness of the sensors to the aggregation, selection, and pre-interpretation of data, all the way to the exact structure of services and business models. On each level, a large number of technologies and systems, social processes and decisions, as well as emergence effects of socio-technical complexity are involved in transforming the perception of the world.

Differentiating between actual information on the world and technological artefacts is essential to successful hermeneutic relations as it guides the way in which output is interpreted. However, this process is rendered impossible by the complexity of mediation. In a few cases, drawing a comparison to naked perception is still possible. Here, the transformation quality might be determined using the difference between naked and machine mediated perception. For example, to monitor the system's answer to the question regarding outside temperature, the user could step outside to feel the temperature or measure it using less complex instruments, that can be interpreted in the usual way, such as a thermometer. However, in the majority of cases, avoiding machine mediation correctively is not possible. For instance, if users are speaking to a single device in their kitchen they are interacting with an extensive network of the *technosphere*. The fact that this interaction is presented as a form of bilateral communication based on a human assistant conceals this network's involvement. It also transforms machine perception from an opaque part of the *technosphere* and a link to numerous human actors that are also part of that *technosphere* to a seemingly pleasant conversation with a personal companion. In this second example, we can observe hermeneutic human-machine-world relations featuring a complex network of *background* relations in the *technosphere*. Here, transformation extent is extremely large, transparency on the machine side of the relation (in the sense of *background* relations being visible and open to human inspection) is almost nonexistent. Due to this opacity, the user lacks important information to be able to interpret the section of the *technosphere* transmitted by the assistance system. The system may present itself as a technical quasi-counterpart that somehow presents some section of the world the user has to interpret in a hermeneutic relation. However, the *technosphere*'s underlying complexity is increasingly putting too much strain on the users' interpreting capacity.

3   Please Don't Let Me Be Misunderstood

In the wake of increasingly personalised systems that feature more and more sensors and that are capable of machine learning using more and more data, also known as big data, the users' interpreting capacity is decreasing while the systems' interpreting capacity keeps increasing. Personalized systems capable of learning are progressively reading users' input in relation to their data profile—their *data shadows* so to speak—meaning the accumulation of past interactions, preferences, and requests of the users themselves, of profiles similar to theirs—their *data twins* so to speak—and the collective body of all users of a learning system, for example all Google Home users. That way, Google Home's multi-user support can differentiate between users via voice recognition. In a family scenario, when asking the question "What is my schedule for tomorrow?", the mother will receive a different output than her son for example. The system does not only interpret the input relative to the speaker but also relative to their data shadows. This phenomenon of machines using information technology to contextualize and relativize humans, inverts Ihde's hermeneutic relation in a way. As a result, we see the world the way in which second type assistance systems present it to us. Information, services, and possible decisions, actions, and interactions are increasingly opaque and can hardly be related back to our naked perception. They have become erratic, contingent and yet somehow necessary in the modality of a deistic god.

The developers and marketing channels of Amazon, Apple, Microsoft, Google etc. do not cease to praise the fact that different assistance products can relieve us of all sorts of burdens. However, in science, politics, art and culture, and in our society, we need to create a dialogue in which we address the possible burdens imposed on us by a permanently and comprehensively assisted lifestyle. In terms of advanced assistive systems, the hermeneutic relation is the dominant relation type, but the vast background relations of the *technosphere* switch the positions of who understands and who is understood. In this brave assisted life those giving up the burden of being on the understanding end of the hermeneutic relation between human and machine-world may find themselves on the opposite end of it. Those giving up the burden of making sense of complex matters may end up being subject to technical capture and interpretation themselves. To be fair, in a hermeneutic relation the position of the *interpreted* is much more comfortable than the position of the *interpreter*. But to cease at least trying to understand the machine-world relations and the *technosphere* background relations personally—even if one might never succeed in doing so—is not an option. Because, when confronted with machine interpretation and technical world mediation where naked perception is not available, the machine mediation has to be compared to something else than naked perception in order to either accept or reject it. Those unburdened by assistance risk to abandon their position as understanding beings in hermeneutic relations to such an extent, that they become assisted as a customer or user, as a data shadow or data source, as someone who is supported or nudged, lead or mislead, understood or misunderstood, but definitely not as someone who understands.

Reference

[Ihd79]   Ihde, D. (1979). Technics and praxis. 1. publ. Dordrecht: Reidel (Synthese library, 130).

Part II
Constructing and Construing

"Construction" refers to the conception and concrete assembly of technical devices. The very process of constructing depends on engineering skill that unfolds via trial-and-error processes in the first place. Recently, these processes have been accelerated to a high degree by, e.g., the possibilities of simulation software or additive manufacturing. However, both beforehand and in the course of the material construction, many questions emerge that beg for answers in form of written or verbal explanations, presentations, or justifications. Hence, material construction is essentially accompanied by a construction of meaning and collaborative relations. That is, any construction depends on construing permanently what is happening, how it might be improved, and what its consequences might be.

This subsection assembles articles that reflect the conditions of construction and present classificatory, philosophical, psychological, and sensorimotor knowledge that is necessary for constructing and construing technology. In general, construction is not some episode during technology development that is completed before subsequent research and testing can take over. Like the analysis of demands and expectations of part I, it is an ongoing process. This iterative and circular character of construction is an effect of the purpose to develop technology that supports human beings. Technology is not only developed *for* people but also *with* the relevant people. This results in a necessity to pass through many iterations.

Construing does also concern the construction process itself. Engineers, but also all other involved developers, interpret the construction process and their role in it persistently. Becoming aware of one's own assumptions during material construction is therefore a prerequisite for making a difference and for engaging in a form of engineering that is intended to develop with people for people. This includes knowledge about possible instruments that help to integrate the different disciplinary backgrounds to achieve a common direction.

This section begins with an article of *Robert Weidner* and *Athanasios Karafillidis* who present a generic view on support systems and offer a classification procedure based on a general theory of support. Classifications work like a backbone for both constructing and construing. They enable the developers to assess and compare their planned or built technical constructions. The authors expound the possibilities of proper classifications and then demonstrate how their proposed classification procedure can be applied to get a classification for exoskeletons.

Janina Loh (née Sombetzki) presents an account of how the construction of autonomous robots (including self-driving cars) can be guided by the idea of responsibility from the outset. After dissecting the concept of responsibility to provide leverage points for deciding whether and when technical systems can be held responsible, she presents the notion of "responsibility networks" to highlight the distributed character of responsibility in sociotechnical arrangements. Due to this, she argues, attempts to attribute responsibility to individual entities are practically not feasible in support systems.

Psychologically informed construction/design choices are the subject of *Rebecca Wiczorek*. She focuses on older people and gives a detailed account of diverse psychological principles and how they can be considered to be able to construct suitable products for older people. Smartphone design is used as a demonstration. Since, according to Wiczorek, the technical implementation of these principles leads sometimes to contradictory design features, construction has to select the aspects relevant for the problem at hand to find a trade-off in each case.

Bettina Wollesen, Laura L. Bischoff, Johannes Rönnfeldt, and *Klaus Mattes* shift our attention to sensorimotor coordination and its impact on designing human–machine interfaces. The authors present different models that explain how various forms of multitasking consume individual cognitive resources in different degrees. They also introduce the concept of "situation awareness" that can push design/construction of relevant and intuitive interfaces in a direction to consume less cognitive resources.

One of the main competences for constructing robots that move around in everyday worlds is, in the words of *Andreas Bischof*, self-awareness of the developers. Robots face "wicked" problems in the wild but developers still ground their definitions of social problems on everyday knowledge that is taken for granted and thus ignores its fundamental ambivalence. Unearthing the underlying assumptions of the very construction of problems that the robots are programmed to solve can add enormous leverage for building robots that people eventually want.

Since the teams constructing support technologies are heterogeneous and display diverse scientific disciplines, construction processes profit from a provision of tools that facilitate diplomacy between them. This is the argument of *Peter Müller* and *Jan-Hendrik Passoth*. In this last piece of part II, the authors draw on experiences from an interdisciplinary collaboration of engineers and social scientists and contend that social scientific methods can be used as diplomatic devices. Also, it is possible to intervene sociologically in a constructive way by selecting a few theoretical ideas as a basis for recommendations during the project.

The empirical process of actual material construction is hardly reflected in engineering. Yet it is incessantly construed due to its equivocality. Knowing more about the underlying premises that guide construction processes adds further possibilities for managing issues or finding alternative project trajectories.

Distinguishing Support Technologies. A General Scheme and Its Application to Exoskeletons

Robert Weidner and Athanasios Karafillidis

Abstract There is a great variety of manual activities in work and everyday life. The number and forms of various support systems reflect this variety. A classification could provide a scheme to recognize, compare, and evaluate the heterogeneous problems, approaches, and technological solutions already existing in the field. Based on previous work on a theory of support, taxonomic criteria are derived and substantiated. These criteria are used to construct a matrix in analogy to the periodic table in order to make visible the coherence as well as the differences of technical support artifacts in general and exoskeletons in particular.

1 Introduction

In recent years many different technical systems have been developed with the purpose to support people in diverse contexts of application. Relevant support technology has been built in particular for the fields of industrial production, construction, elderly care, and also for daily life, e.g., mobility solutions, smart homes, or health applications. Exemplary support systems range from industrial and service robots to implants. Located somewhere in between one finds a plethora of various other technical solutions, for example lifting aids, optical aids, electro bikes, apps for mobile devices, tools, assistance systems for cars, airplanes, or industrial production, and exoskeletons for rehabilitation, agriculture, military, or assembly tasks. All of these systems are structurally different. They are based on different approaches and are obviously characterized by different interaction patterns of their heteroge-

R. Weidner (✉) · A. Karafillidis
Laboratory of Manufacturing Technology, Helmut Schmidt University/University of the Federal Armed Forces Hamburg, Holstenhofweg 85, 22043 Hamburg, Germany
e-mail: robert.weidner@hsu-hh.de

A. Karafillidis
e-mail: karafillidis@hsu-hh.de

R. Weidner
Chair of Production Technology, University of Innsbruck, Innsbruck, Austria

© Springer Nature Switzerland AG 2018
A. Karafillidis and R. Weidner (eds.), *Developing Support Technologies*, Biosystems & Biorobotics 23, https://doi.org/10.1007/978-3-030-01836-8_8

neous components. Yet all of them are developed and seen as supporting people and their activities.

To develop technology with supportive purposes is not just a trend that will fade away as trends by definition always do. It is rather a new frame for constructing and construing the relationship between humans and technology. The intention to build devices that *support* people recasts the whole development process of technology. Thus, new conceptual approaches for comparing and evaluating such technologies are required.

In what follows, one such basic approach is presented. Based on previous theoretical considerations [Wei15, Kar15, Wei16, Kar16, Kar17] basic taxonomic criteria for support systems are stipulated and a classification procedure described. In order to improve the understandability, exemplary systems will be used for explanation and classification. The resulting taxonomy for support systems in general and exoskeletons in particular is going to exploit the idea of the periodic table. Using this form for classification has several advantages. First of all, it helps to compare existing systems, which is of great importance for assessing them but also for selecting suitable systems and components. Then it is expected to enable developers to devise and design future support systems efficiently according to defined requirements as well as social, ethical, and legal contexts and obligations. Finally, such a systematic will help to identify gaps that indicate possible technical support solutions not yet invented.

2 Support Systems

A support system is not simply a technical device but rather the result of a succeeding connectivity between humans (i.e., bodies, perceptions, thoughts), a technical device (i.e., electro-mechanical structures, eventually sensors and actuators), and activities (i.e., situations, expectations, bio-physical environment). Any technological invention has to take this into account to keep its chance of being accepted. Since these technical artefacts are usually understood to represent the system as a whole, the following taxonomical considerations will be related to these technical devices proper—with their names and their presumably clear-cut boundaries.

The term "support" has a particular function in the research networks of relevant technology. It does not only refer to the intended technical functioning but also frames the social-cultural understanding and justification of such technologies, for example, by demonstrating their logical necessity or by giving legitimate reasons for their deployment. Such demonstrations and reasons for support mostly address benefits for people but also for organizations or society in general. Prominent among them are promises of increasing productivity and product quality as well as decreasing physical and cognitive stress. With regard to the activity, support systems are seen as providing facilitation. From the perspective of an individual, support technologies are expected to provide *relief*. It sounds promising and fascinating to have technical systems that support people and their activities when bodily movements, complex situations, procedures, or forms of collaboration become too demanding.

The technical solutions for individual needs are mainly justified by this search for relief, which is assumed to be a general human aspiration. Relief can be provided in different respects. Thus, engineering research for solutions that yield individual relief evolved in separate, yet overlapping fields, that are concerned with either physical or psychological, or processual, or "social" relief. A summary is shown in the following Fig. 1.

The distinction of these fields is a somewhat refined version of the more common distinction between physical and cognitive support of human capabilities, which harks back to the age-old Cartesian separation between mind and body. In the wake of embedded systems [Hen07] and distributed cognition [Hut95] this distinction is obviously dated and skewed, but it does still lead most attempts to find suitable technological leverage points and classifications for support technologies.

Viewed in this refined, yet still classical scheme, this paper will focus on technical support for physical relief: exoskeletons. However, it will not proceed by sticking to these distinctions. The aim is to introduce a new pattern for getting observations of support technologies and an ensuing classificatory procedure, that cuts across these classical fields. For both technology development and research in technology it is necessary to find out how these various fields depicted in Fig. 1 are connected. This marks the route that leads from simple technological inventions to their societal acceptance, that is: to innovation [Ram07].

The subsequent paragraphs will introduce a classification procedure, which is preoccupied with *relations* between humans, technology, and situations. The main determinants that shape such support relations and result in the stability and reliability of a support *system* are temporality, coupling, and control. Their deduction has been described in detail elsewhere [Kar16, Kar17]. Any technological device from an industrial robot to a body implant displays particular decisions with respect to these determinants. After setting up the general procedure along the lines of the periodic table it is applied to existing exoskeletal solutions.

kind of support			
physical relief	**psychological relief**	**organizational relief**	**social relief**

- force redirection
- force induction
- force enhancement
- stabilization
- bracing
- ergonomic
  improvements
- precision
	- work instruction
	- activity monitoring
	- stress monitoring
	- compensation of
	  feelings
	- time management
	- distribution of
	  activities
	- job security
		- work instructions
		- monitoring of process
		- monitoring of working
		  sequences
		- monitoring of system
		  states
		- suggestions and
		  improvements
			- facilitating contacts
			- flow of
			  communication
			- distribution of
			  activities
			- smoothing
			  hierarchical
			  differences
			- team work
			- encouragement
			- knowledge sharing

Fig. 1. Promises/expectations of technical support with examples [Wei16]

3   Classification with a Periodic Table

Classifying is both an everyday practice and a scientific procedure [Lev73]. Scientific classifications mark the distinctions used to classify explicitly. They become complete when they do not only organize knowledge but also specify ignorance. A scientifically apt classification should do both, but only few methods have a keen eye for the gaps of a classification and recognize their importance for research. One of these methods is the periodic table. When Mendeleev published his ideas in 1871 he arranged the known elements such that their relations became visible (organizing knowledge) and that he could predict "missing" elements not yet discovered (specifying ignorance).

There is an issue involved in using the periodic table for classification. It has not been recognized to be a general classification method but as being confined to the discipline of chemistry. However, a suitable generalization is not difficult to achieve. The construction principle of Mendeleev's original periodic table of the elements is very straightforward [Bac80]. Following his original lead, two dimensions have to be found to get a kind of matrix-structure.

Before, two questions have to be answered, like in any other classification procedure: (a) What is to be classified? and (b) What constitutes an "element"? The first question addresses the population (e.g., flowers, vertebrates, chemical elements, technical devices). Technically, the answer to this question already determines the answer to the second one. But since there is no "natural" baseline of what constitutes elements it depends on the observing agents whether single entities, or parts of them or groups of them are considered as elementary for the classification. Therefore, it makes sense to distinguish these two questions.

3.1   Population: Technical Artefacts for Support

Support is a very far reaching and diverse phenomenon. Instead of classifying support in general, we confine the classification to technical artefacts that are deployed to support human activities. All artefacts observed as support and assistance systems are part of the empirical population. The objective is to make heterogeneous technical solutions comparable and to render their relations visible. Despite this strong limitation, the variety remains high. Such a generalized approach has not been attempted so far. Domain-specific classification approaches for software [Dog12], operator information systems [Teu16], or exoskeletons [Hoc15] are preferred. The periodic classification procedure can also be applied to such narrower populations of support systems. This will be exemplified for the case of exoskeletons.

3.2 Empirical Elements: Deployed Technical Support Solutions

The elements to be classified will not be specific, individual technical solutions of certain manufacturers or research institutes. Their numbers are too high to achieve a viable classification. This would be a proper task for numerical taxonomies, which could be used to complement a periodic table [Sne73]. The field of support systems is very dynamic, so that any classification attempt applied at the level of individual technical objects would be problematic anyway. The elements of the population are thus assumed to have a minimal structure, which is already present in this field. Classified are for example industrial robots, exoskeletons, or autonomous vehicles and not some specific industrial robot or particular vehicle of manufacturer XYZ. In the next step it is possible to focus on, e.g., exoskeletons to refine the observations of these particular systems and their differences and relations.

3.3 Taxonomic Dimensions: Communicative Patterns and the Perception of Qualities

For developing the periodic table, two dimensions are needed to get its matrix form. The first short periodic table of Mendeleev constituted the rows with reference to the periodic "law", that is, it displayed a sequential ordering of elements by their atomic number and their repeatedly discernable properties. The columns represented groups of elements pooled with regard to their similar physical and chemical qualities, e.g., their oxidation states [Sce11].

The generalized version presented here starts from situations of support. In sociology, situations are something like an always moving, dynamic, and indeterminate experimental set-up in which for example different reactants, connections, substances, and catalysts (i.e., objects, people, times, and relations) are combined and recombined to reproduce certain patterns or bring forth new ones. Any situation performs and accomplishes in some form *society*, that is, it tests the association of different and independent entities. Society is not an overarching whole, but rather a process of looking for viable ways of association by finding an alignment between communication and perception [Luh12]. Therefore, the rows of the generalized periodic classification show different communication patterns that arise in every support situation and the columns sum up the perception of the qualities of technical systems ascribed and inscribed to them with reference to the human body.

3.3.1 Communicative Patterns of Support Situations (Ordinate)

For the classification of support systems, three central determinants have been identified [Wei15, Kar15, Wei16, Kar16, Kar17] that determine the support situation and the overall context. Since the focus is on situations neither object-specific properties, which are related to the material and structural nature of artifacts (e.g., actuators, sensors, kinematics, material, or structure), nor subject-specific factors, which usually distinguish cognitive and physical functions are used as starting points for support. We rather look for situational relations between artifacts and humans.

In situations of support a supported activity is distinguished from a support activity. Mostly—at least within the realm of technology development, that will guide the following considerations—a distinction is made between a supported human activity (this may be an "internal" activity, such as physical functions, or an "external" task, such as drilling a hole) and the technical artifact that is meant to support this human activity. Their communicative relationship is determined by three factors: (a) their temporal relation, (b) the form of their coupling, and (c) the attribution of control, see the schematic representation in Fig. 2.

Temporal Relation (between Human and Technology)

The relation between human activity and support technology runs either synchronized or desynchronized. Synchronization is possible and observable, when there is a reciprocal interaction between human and technology. This usually implies that the technical support is performed within the boundary of the situation. If on the other hand technology and human beings do not sense the respective activity of the other in either direction (e.g., a software runs in the background, without reacting immediately to current human activities) the criterion of reciprocity is not met, and synchronization does not take place—even if an observer can possibly describe the relation as synchronized in some form. The necessity of mutual perception for activities to be synchronized is not limited to visual sensation. If there is no synchronization, the relation is not co-present, but distributed over time and place.

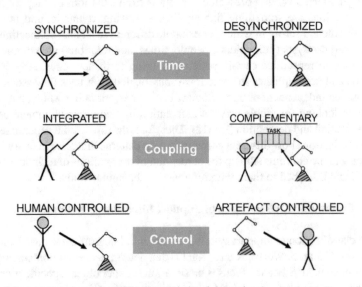

Fig. 2. Schematic representation of the three relational determinants

Relation of Coupling (between Human and Technology)

Human and technological activities can be integrated or complementary. Integration is the case, when the technical support partakes in the execution of the activity operationally and materially. The degrees of freedom of the technology are then bound and constrained by the degrees of freedom of the human activity. The integration of activities in most cases targets the support of human functions and their compensation, recovery, replacement, or reinforcement. In contrast, a complementary coupling exists when the relation between human and technology is collaborative. Such a division of labor becomes possible when the activities necessary for task accomplishment can be subdivided into different steps or components. This includes the contexts of the activity, too. In the exact sense, the term "assistance" that is commonly used to indicate technical support systems refers only to this complementary form of coupling. Hence, assistance indicates a special form of support that is characterized by a division of labor between human and technology.

Perceived Control Relation (between Human and Technology)

Observers involved in support situations ascribe control either to human beings or to technology. In general, control is not a one-sided affair but circular, distributed, and depending on communication [Gla87, Ash58, Vic67, Whi08]: to exert control requires to allow one's own activities to be controlled by the controlled entity. Human beings control the technology and, for example, allow technology to control their movements or their field of vision, which triggers further control attempts in return—and so forth. However, this circularity always becomes punctuated [Wat00], that is, an origin of control is ascertained during the process. Control is attributed to one of the participating entities. This perception of control might differ from the control intentions of the developers, it can frame the control relation as permanent or temporary, and it might also be seen to change in time or to oscillate within the current support situation.

For simplicity, these determinants are understood dichotomously. This results in 2^3 possible communication patterns of support. Their deduction is shown in the following Table 1.

A "1" indicates that the respective structural expression is present. A "0" indicates that it is absent. There may be situations/artefacts that realize both sides of a determinant simultaneously or oscillate between two determinants, but these cases are not considered here.

The three-letter codes on the left of the table serve as abbreviations for the eight different relational patterns in support situations (e.g. DCA denotes **D**esynchronized, **C**omplementary, and **A**rtefact-controlled relational pattern). They are clustered into two groups, the desynchronous group (D-group, i.e. all codes beginning with D; first four rows in the table) and the synchronous group (S-group, i.e. all codes beginning with S; last four rows in the table). Such clustering and possible regroupings are of great help for a comparison of the technical solutions in the resulting classification.

Table 1 Basic communication patterns of support situations

	Temporal relation		Coupling		Control attribution	
	Desynchronized	Synchronized	Complementary	Integrated	Artefact-controlled	Human controlled
DCA	1	0	1	0	1	0
DCH	1	0	1	0	0	1
DIH	1	0	0	1	0	1
DIA	1	0	0	1	1	0
SCH	0	1	1	0	0	1
SCA	0	1	1	0	1	0
SIA	0	1	0	1	1	0
SIH	0	1	0	1	0	1

3.3.2 Perception of Artefacts in Relation to the Human Body (Abscissa)

The eight mentioned patterns of communicative relations, define one axis of a two-dimensional matrix constructed in analogy to the periodic system. These are recurring patterns of support situations. The other axis of the matrix groups the technical systems according to their perceived similarity with respect to size and distance to the human body.

Five ratios of bodies and technical artefacts are distinguishable though not clearly distinct in all respects. Some technical systems cannot be assigned to one ratio unambiguously. Like all the ideas and distinctions presented here this is contingent on the observers' interests and the context of application. However, artefacts that share a particular ratio are considered similar.

Separated

Artefacts with a body of their own (not with simple housings). They operate either autonomously or are operated by remote. Therefore, they can vary with respect to spatial proximity and distance to the human body. Examples are robots of all kinds, drones, and systems for telemanipulation.

Ambient

Artefacts whose embodiment is distributed and whose unit is therefore not directly perceptible. The identity is sometimes condensed in (often mobile) interface devices, which make their technological context visible and understandable. Examples are Ambient Assisted Living and smart homes. Autonomous vehicles are somewhere in between separate and ambient support technology.

Close to the Body

Small up to medium-sized devices/objects with which humans relate either by rudimentary interaction or by operating them. Included are mobile phones, but also classic tools, or lifting aids. Close to the body refers to the grasping radius or to the variable territory of self [Gof71]. The combination "both ambient and close to the body" can be found in many technologies, e.g., driver assistance systems.

Wearable

Artefacts that are wearable or fixed directly on the body (this does not include for example objects somebody transports inside a pocket or in the hand). They have the maximum size of human bodies. The minimum size has in principle no limits. Examples are fitness bracelets, hearing aids, simple glasses, and exoskeletons.

Implanted

Artefacts that are implanted into the body via various pathways (mostly operative, but also by swallowing). They are correspondingly connected to internal organic structures (often strands of nerves) or do temporarily observe physical processes/parameters. Examples are pacemakers, electrodes for neurostimulation, and data pills. A limiting case are portable and implanted neuromuscularly controlled arm prostheses.

3.4 Periodic Classification Table

Putting together the two dimensions of the generalized periodic table, a qualitative, empirical classification of technical support systems can be carried out. We chose this two-dimensional representation, because it allows to visualize elementary relations between different technical systems. The general approach will be complemented by an approach for exoskeletons—with a different level of detail.

Elementary and recurring relational patterns of support situations are entered on the vertical axis—and perceived body-artefact-relations in the horizontal axis. A "periodic law" is missing but even in the case of the chemical periodic table, the periodic law was not decisive for the arrangement of the elements. At first, purely qualitative criteria obtained—the qualitatively perceived similarity of the chemical properties of individual substances [Sce12].

Similar to the first version of Mendeleev—a 12×8 table [Sce11]—we get an 8×5 table in which certain solutions are grouped (see Fig. 3). The horizontal axis consists of five columns: separated, ambient, close to the body, wearable, and implanted. The "width" of the five columns has no particular meaning in this version. The chosen sequence within the rows follows simply the size of the artefact relative to the body. In future, this differentiation can be used to distinguish further between technical

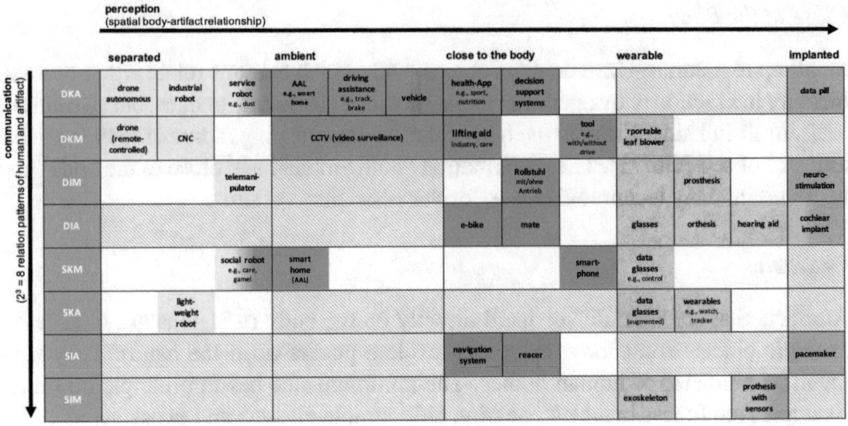

	separated		ambient				close to the body		wearable				implanted
DKA	drone autonomous	industrial robot	service robot e.g. dust	AAL e.g. smart home	driving assistance e.g. track, brake	vehicle	health-App e.g. sport, nutrition	decision support systems					data pill
DKM	drone (remote-controlled)	CNC		CCTV (video surveillance)			lifting aid industry, care		tool e.g. with/without drive	rportable leaf blower			
DIM			telemanipulator					Rollstuhl mit/ohne Antrieb			prosthesis		neuro-stimulation
DIA							e-bike	mate		glasses	orthesis	hearing aid	cochlear implant
SKM			social robot e.g. care, gamel	smart home (AAL)					smart-phone	data glasses e.g. control			
SKA		light-weight robot								data glasses (augmented)	wearables e.g. watch, tracker		
SIA							navigation system	reacer					pacemaker
SIM										exoskeleton	prothesis with sensors		

perception (spatial body-artifact relationship)

communication (2³ = 8 relation patterns of human and artifact)

Fig. 3. "Periodic system" for a classification of technical support systems

systems. Likewise, basic values could be entered into the respective boxes to further refine the differentiation.

The process of identifying elements for certain classes will not be described in detail, albeit this could probably be the cause for some fundamental questions (Why is the technology X in cell Y?). At this point, it is sufficient to accept that the identification has been conducted based on the previously introduced criteria. Allocation disputes will thus be spared—there is no unique position for each item. This classification cannot hide its observer dependence anyway.

A first inspection of the table reveals three distinctive features:

(1) Some similarities were to be expected, others are surprising. The fact, that an industrial robot and a service robot are closer to each other than a smart home and a hearing aid is not surprising. The fact, that heart pacemakers and navigation systems or e-bikes and hearing aids have similar relation patterns (they can be found in one row, namely the SIA or DIA pattern), is astonishing though. It is also interesting to note that technology might switch to other patterns (as is illustrated in the table with drones, data glasses, and exoskeletons) when further developments or different types of usage are added—for example, the integration of sensors or the use of data glasses only for taking photos.

(2) The body-centered and wearable artefacts show the greatest variance with respect to the relational pattern of support—most of the cells in these two columns, six of eight, are occupied. The lowest variance in this respect is currently found with ambient technologies. Furthermore, the largest variation in technical support forms is found in the relational pattern DKA (Desynchronized, Complementary, Artefact-controlled). This is historically important. We can see a shift from the D-group (the first four lines) to the S-group (the lower four lines) when moving from left to right in the table, that is, from bigger to smaller artefacts and closer to the body.

(3) The entire matrix is slightly more than half filled (24 of 40 cells). Almost half of it shows vacancies (shaded in the matrix). These vacancies are not a problem of this classification, but rather one of its most important functions, because they indicate potential future support technologies.

4 A Periodic Classification of Exoskeletons

This periodic classification for technical support systems can be used to dig deeper into the diversity of technical solutions in particular fields. Any of the cells can be cracked open to discover the variety of the technological category classified within the general scheme. In a first approximation, exoskeletal systems will serve here as a trial.

A direct comparison of system solutions, e.g., two different exoskeletons, requires a further refinement of the proposed classification procedure. Criteria like weight, materials used, costs, or flexibility with respect to human anthropometry are not of interest for the classification. The criteria should rather be refined to address the technical characteristics.

A subdivision that is widespread in the literature concerns the power. Here, a distinction is made between active (external sources like motor or battery), passive (not powered by external sources; e.g., work on mechanical linkages, pneumatic and hydraulic mechanisms, or springs), and hybrid systems. Moreover, exoskeletons are often classified in respect to the supported body area (e.g., lower extremities, upper extremities, hand and back) and domain of application (e.g., rehabilitation, agriculture, military, and industrial production). In what follows, we stick to the above discovered relational communication patterns between human and exoskeleton as one axis and adapt it accordingly. The second axis displays the perception of the supported area of the body. This perception does mainly refer to the perceptions and beliefs of the developers and stakeholders about which body parts should be supported—simply because this technology is only just beginning to leave the labs.

4.1 Relational Patterns of Human and Exoskeleton

Our first examination with the usual categories of exoskeletons and the above developed patterns has identified five central patterns between human and exoskeleton. At this stage of research, we did not look for a perfect match between them. However, this step is mandatory if the aim is to achieve a more robust classification that exploits all the possibilities of the periodic classification.

In the context of wearable robotics, the identified relational patterns refer to the form of interaction between human body *movements* and the exoskeletal solution realized during support situations. Support may refer to, for example, the amplification of movements, the following of movements, or the stabilization of postures. The following patterns are used for the classification.

Enabling Movements

Exoskeletal systems which enable a certain movement that cannot be performed (anymore). The relevant relational pattern is mostly integrated, that is, the exoskeleton is constitutive of the movement in a very radical way. The control is in most cases effected by another person, e.g., a therapist, or by measuring cerebral activity or nerves of other body parts. Examples are exoskeletons for cross-section paralysis.

Empowering Movements

Exoskeletal systems that support human movements that are still within the reach of the human performer. This implies that natural functionality is available and the system, for example, increases force. The support must be felt significantly. It empowers users to perform tasks which they cannot perform without support. Therefore, such systems are seen to control the support situation. Examples are tasks which need more maximum force than the user can apply.

Facilitating Movements

Exoskeletal systems that do not obstruct natural human movements but support in defined situations in such a way, that, for example, the risk of overload is minimized. Compared to the empowering of movements these systems do not allow the execution of tasks beyond human capabilities. They rather improve ergonomic conditions and thus facilitate movements. An example here is the absorption of overload during specific subtasks. Relevant systems enforce the deployment of human functionality. Human body movements are neither substituted nor complemented but rather integrated and strengthened. Muscles are activated within a range considered as healthy and the exoskeleton reduces musculoskeletal stress that exceeds some (individually adjustable) limit.

Stabilizing Movements

The fourth pattern can be observed when systems are used to stabilize one or more body parts or human joints during tasks. The support relation is more static in nature than in the other patterns described this far. Thus, such systems do not support dynamic but static movements. Holding a posture is still a movement. First, it requires micro-movements (postural control) to stand still and second, the stabilization of a posture is also only part of a more comprehensive movement process. External loads can be partially or completely removed with this kind of exoskeletons. Examples are back support during tasks like the handling of heavy objects or a support of the shoulder and upper extremities during tasks at and above head level.

Table 2 Relation between interaction patterns and basic patterns of support situations

	Temporal relation		Coupling		Control attribution	
	Desyn-chronized	Synchron-ized	Comple-mentary	Integrated	Artefact-controlled	Human controlled
Enabling Movements	1	0	0	1	0	1
Empowering Movements	0	1	0	1	1	1
Facilitating Movements	1	1	0	1	0	1
Stabilizing Movements	1	1	1	1	1	0
Adding Movements	1	1	1	0	0	1

Adding Movements

The fifth pattern of relations between exoskeleton and human body movement refers to systems for explicit tool support, for example during handling tasks by third arm solutions. This adds movement possibilities to the human body. Relevant systems are characterized by the fact that they have a direct connection (possibly also detachable) to the tool and are arranged parallel to at least one body part, but desynchronized and without direct coupling. These could be the only exoskeletal solutions available right now that are designed in a complementary fashion, that is, in which the task is divided into subtasks that are handled based on a division of labor between exoskeleton and human being.

Assigning the described relational patterns to the basic patterns of support situations described above results in the following table (Table 2).

It is clearly visible, that not all of the eight possible patterns developed in Table 1 are realized. Additionally, it is conspicuous that in some patterns two opposing features are both present, for example they are synchronized and desynchronized at the same time. That is because different solutions exist for the one and the other feature and both are crammed into one pattern (row). At this juncture of the periodic classification for exoskeletons, further differentiation of the described patterns is unquestionably possible and necessary.

4.2   Supported Movements of Body Parts

The support provided by exoskeletons addresses different body parts, for example, ankle, knee, hip, lower back, upper back, shoulder, elbow, wrist, finger, or neck. Full body exoskeletons are not viable at the moment but are also considered for reasons of completeness and also because this could change in future.

The technical solutions can be focused solely on one specific part of the body or may involve a combination of several parts. As a rule, the signals (insofar sensors are used) or the movement or poses are detected directly at the body part that is to be supported. However, it is also possible that the set point/target is picked up elsewhere, for example, by measurement of brain currents. But these cases are not considered here in detail.

4.3 A Tentative Periodic Table for Exoskeletons

The following periodic table is a result of the previous considerations. It displays exemplary, but still in most cases generic solutions (Fig. 4).

This form of classification is not valid once-and-for-all. On the one hand it is tentative with respect to its inherent possibilities that have not been exploited yet. On the other hand, it is of utmost importance to comprehend that the periodic classification is a *procedure* in the first place. It has to be implemented and reimplemented depending on the systems of interest, the domain of application, and the continuing progress in the field of exoskeletal solutions.

At first glance the periodic table of exoskeletons shows some interesting points. The vacancies are of major importance because they indicate that either with regard to particular body parts or to relational patterns of interaction some solutions are missing. This is not to say, that any vacancy hast to be filled. But it might give hints for future demands and technical possibilities. Additionally, heterogenous and diverse systems might now be compared (e.g., muscle gloves and lower extremities exoskeletons) because they realize the same pattern. Finally, such comparison may be used to discover technical solutions that have already been found for other exoskeletons that have not yet been considered as similar at all.

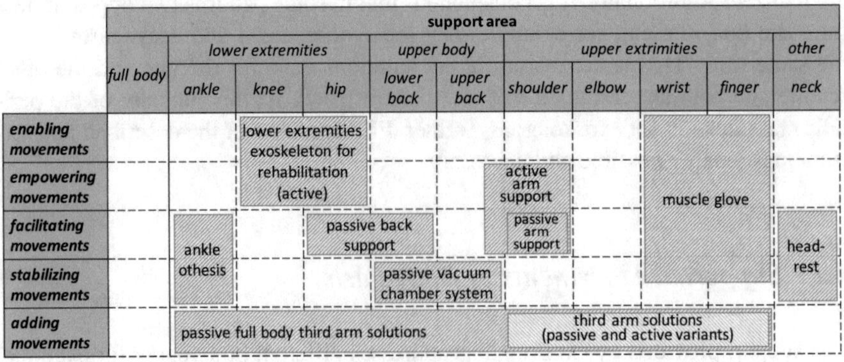

		support area									
	full body	lower extremities			upper body		upper extremities				other
		ankle	knee	hip	lower back	upper back	shoulder	elbow	wrist	finger	neck
enabling movements			lower extremities exoskeleton for rehabilitation (active)						muscle glove		
empowering movements							active arm support				
facilitating movements		ankle othesis		passive back support			passive arm support				head-rest
stabilizing movements					passive vacuum chamber system						
adding movements		passive full body third arm solutions					third arm solutions (passive and active variants)				

Fig. 4. Classification of some exemplary solutions from state of the art

5 Conclusion

Numerous systems to support people in different contexts do already exist. Further solutions are emerging. The presented periodic classification procedure leaves inappropriate and gridlocked classical distinctions like physical/cognitive support behind and provides a fresh perspective and some orientation. Besides, we demonstrated its flexible applicability by shifting the degree of detail: From a general periodic table for technical support systems to one for exoskeletons.

The introduced procedure can be used for any systematic classification in the heterogeneous field of technical support systems. At its core are the determinants and criteria that have been developed with a focus on interactional patterns in support situations. Starting from there, both surprising and obvious relationships and differences between technical objects can be identified.

The resulting periodic tables allow to identify gaps in research and development, to indicate problems more precisely, and to specify our ignorance in this field of technical support systems. Such a form of classification defies the many unsystematic and never really explicable classifications that are mostly based on nothing more than everyday experience of developers and stakeholders. It has a theoretical basis in sociological theory and is a mixture of empirical and conceptual approaches. It is neither simply an (empirical) taxonomy of existing systems nor simply a (conceptual) typology, but a kind of "taxology", or even better: a periodic classification.

Acknowledgements This research is part of the project "smartASSIST—Smart, AdjuStable, Soft and Intelligent Support Technologies" funded by the German Federal Ministry of Education and Research (BMBF, funding No. 16SV7114) and supervised by VDI/VDE Innovation + Technik GmbH.

References

[Ash58]   Ashby, W. R. (1958). Requisite variety and its implications for the control of complex systems. *Cybernetica, 1*(2), 83–99.

[Bac80]   Bachelard, G. (1980). *Philosophie des Nein*. Frankfurt a. M.: Suhrkamp.

[Dog12]   Dogangün, A. (2012). *Adaptive awareness-assistenten: Entwicklung und empirische Untersuchung der Wirksamkeit*. Lohmar/Köln: Eul Verlag.

[Gla87]   Glanville, R. (1987). The question of cybernetics. *Cybernetics and Systems, 18*(2), 99–112.

[Gof71]   Goffman, E. (1971). Relations in public. Microstudies of the public order. New York: Harper & Row.

[Hen07]   Henzinger, T. A. & Sifakis, J. (2007). The discipline of embedded systems design. Computer 10, *IEEE Computer Society*, S. 32–40.

[Hoc15]   Hochberg, C., Schwarz, O., & Schneider, U. (2015). Aspects of human engineering—bio-optimized design of wearable machines. In A. Verl, A. Albu-Schäffer, O. Brock, & A. Raatz (Eds.), *Soft robotics. Transferring theory to application* (pp. 184–197). Berlin: Springer.

[Hut95]   Hutchins, E. (1995). *Cognition in the wild*. Cambridge: MIT Press.

[Kar15]   Karafillidis, A., & Weidner, R. (2015). Grundlagen einer Theorie und Klassifikation tech-
           nischer Unterstützung. In R. Weidner, T. Redlich, & J. P. Wulfsberg (Eds.), *Technische
           Unterstützungssysteme* (pp. 66–89). Berlin: Springer-Verlag.

[Kar16]   Karafillidis, A.,& Weidner, R. (2016). Taxonomische Kriterien technischer Unter-
           stützung – Auf dem Weg zu einem Periodensystem. In R. Weidner (Ed.), 2. *Trans-
           disziplinäre Konferenz "Technische Unterstützungssysteme, die die Menschen wirklich
           wollen"* (pp. 233–247).

[Kar17]   Karafillidis, A. (2017). Synchronisierung, Kopplung und Kontrolle in Netzwerken. Zur
           sozialen Form von (technischer) Unterstützung und Assistenz. In P. Biniok & E. Lettke-
           mann (Eds.), *Assistive Gesellschaft* (pp. 27–58). Wiesbaden: Springer VS.

[Lev73]   Lévi-Strauss, C. (1973). *Das wilde Denken*. Frankfurt a. M.: Suhrkamp.

[Luh12]   Luhmann, N. (2012). *Theory of society*. Two Volumes. Stanford UP.

[Ram07]   Rammert, W. (2007). *Technik – Handeln – Wissen. Zu einer pragmatistischen Technik-
           und Sozialtheorie*. Wiesbaden: VS Verlag.

[Sce11]   Scerri, E. R. (2011). *The periodic table*. Oxford: Oxford UP.

[Sce12]   Scerri, E. R. (2012). A critique of Weisberg's view on the periodic table and some
           speculations on the nature of classifications. *Foundations of Chemistry, 14*, 275–284.

[Sne73]   Sneath, P. H. A., & Sokal, R. R. (1973). *Numerical taxonomy*. San Francisco: W. H.
           Freeman and Co.

[Teu16]   Teubner, S., Reinhart, G., Haymerle, R., & Merschbecker, U. (2016). Individuelle und
           dynamische Werkerinformationssysteme. In R. Weidner (Ed.), *Band zur zweiten trans-
           disziplinären Konferenz "Technische Unterstützungssysteme, die die Menschen wirklich
           wollen"*, Hamburg (pp. 349–364).

[Vic67]   Vickers, G. (1967). Cybernetics and the management of men. In: ders. (Ed.), *Towards a
           sociology of management* (pp. 15–24). London: Chapman & Hall.

[Wat00]   Watzlawick, P., Beavin, J. H., & Jackson, D. D. (2000). *Menschliche Kommunikation.
           Formen, Störungen: Paradoxien*. Hans Huber, Bern.

[Wei15]   Weidner, R., & Karafillidis, A. (2015). Three general determinants of support-systems.
           In *Applied mechanics and materials* (Vol. 794, pp. 555–562). Schweiz: Trans Tech
           Publications.

[Wei16]   Weidner, R., Karafillidis, A., & Wulfsberg, J. P. (2016). Individual support in industrial
           production—Outline of a theory of support-systems. In *49th Annual Hawaii Interna-
           tional Conference on System Sciences* (pp. 569–579).

[Whi08]   White, H. C. (2008). *Identity and control. How social formations emerge* (2nd ed.)
           Princeton UP.

On Building Responsible Robots

Janina Loh (née Sombetzki)

Abstract Rapid progress in robotics and AI potentially pose huge challenges regarding several roles that used to be traditionally reserved for human agents: Human core competences such as autonomy, agency, and responsibility might one day apply to artificial systems as well. I will give an overview on the philosophical discipline of robot ethics via the phenomenon of responsibility as a crucial human competence. In a first step I will ask for the traditional understanding of the term "responsibility and formulate a minimal definition that exclusively includes the necessary etymological elements as the 'lowest common denominator' of the responsibility concept: Responsibility as the ability to answer is a normative concept that rests on the assumption that the responsible subject in question is equipped with a specific psycho-motivational constitution. In a second step I will outline my understanding of the discipline of robot ethics, in order to ask in a third step how to ascribe responsibility in man-machine-interaction. For these purposes I will elaborate on my concept of responsibility networks.

1 What Is Responsibility?

A detailed etymological study would show that our understanding of "responsibility" rests on three crucial premises [Som14, pp. 33–41]: It means—firstly—"to be answerable for something". It is the ability to answer when someone needs to explain her- or himself. Secondly, responsibility is a normative concept, i.e., it is not only descriptive and causal. In calling the rain responsible for wetting the street we use the term "responsible" in a metaphorical sense because the rain is not able to explain itself. In—on the other hand—calling someone responsible for killing another person we usually do not want to state a simple fact or see the person in question as a cause in a purely descriptive way. We want the claimed murderer to explain her- or him-

J. Loh (née Sombetzki) (✉)
Department of Philosophy, Philosophy of Technology and Media, University of Vienna,
Universitätsstraße 7 (NIG), A-1010 Vienna, Austria
e-mail: janina.loh@univie.ac.at

© Springer Nature Switzerland AG 2018
A. Karafillidis and R. Weidner (eds.), *Developing Support Technologies*, Biosystems &
Biorobotics 23, https://doi.org/10.1007/978-3-030-01836-8_9

101

self and to accept her or his being guilty. Finally, responsibility includes a specific psycho-motivational constitution of the responsible subject in question: We think her to be answerable in the sense of being an autonomous person, to feel addressed to take up her responsibility and to be equipped with several capabilities such as judgment and reflective faculty [Loh17a, Som14, pp. 39–41].

This etymological minimal definition of responsibility leads to five relational elements: An individual or collective subject or bearer of responsibility as the responsible agent or person (the *who* is responsible?). The subject is prospectively or retrospectively responsible for an object or matter (the *what* is x responsible *for*?). The subject is responsible to a private or official authority (the *to whom* is x responsible?) and *towards* a private or official addressee or receiver. The addressee is the reason for speaking of responsibility in the context in question. Finally, the (private or official) normative criteria define the *conditions under which* x is responsible. They restrict the area of responsible acting and by this differentiate moral, political, legal, economic and other responsibilities, or better: domains of responsibility. A thief (= individual subject), is for instance responsible for a stolen book (= retrospective object; better: the theft, a collection of actions that already happened) to the judge (= official authority) towards the owner of the book (= official addressee) under the conditions of the criminal code (= normative criteria that define a legal or criminal responsibility).

In the light of this minimal definition of responsibility it becomes clear that a complex cluster of capacities is needed to call someone responsible: The ability to communicate, autonomy (that includes being aware of the consequences, i.e., knowledge, being aware of the context, i.e., historicity, personhood, and a scope of influence), and judgment (that includes several cognitive capacities such as reflection and rationality as well as interpersonal institutions such as promise, trust, and reliability). It is important to take into consideration that these three sets of capacities can be ascribed in a gradual manner. As it is possible to speak of more or less communication skills, to say that someone is more or less able to act in a specific situation, she is more or less autonomous, reasonable, and so on, it follows that responsibility itself must be attributed gradually according to the present prerequisites. Assigning responsibility is not a question of "all or nothing" but one of degrees.

2   What Is Robot Ethics?

The discipline of robot ethics includes two complementary fields of research: Some thinkers ask whether robots are to be understood as "moral patients (as entities that can be acted upon for good or evil)" [Flo04, p. 349], others concentrate on the question whether robots are themselves "moral agents ([…] entities that can perform actions […])" (ibid.). The group of moral agents is more exclusive compared to the group of moral patients; commonly only humans—and by no means every human being, since, e.g., children and people with disabilities, handicaps, or due to accidents can temporarily as well as generally be excluded from moral agency—are seen as equipped with the necessary capacities for moral agency. Numerous beings and entities such as animals, plants, but things as well (e.g., a car, smartphone, house) have a (instrumental) value and are morally worth considering. A moral agent is likewise a moral patient—but not vice versa. Living creatures and things have a moral value dependent on the underlying ethical approach; for instance, an anthropocentric, pathocentric, biocentric, or physiocentric view.

Within the research field of robots as moral patients the core interest lies in human behavior towards artificial systems. Here, thinkers ask how one should 'treat' robots (as compared to animals and children) and whether they have a moral value even though they might themselves not be able to morally act. As moral patients, artificial systems are without exception understood as tools and supplements to humans. One might for example consider ethical codes for companies and firms, others might concentrate on the question whether relations to and with robots are possible and desirable, whether one could 'enslave' robots, and how to evaluate therapeutic artificial support systems and robots. Within this research area exclusively the human designers and users have the moral competence and competence-competence. The human 'parents' decide on the morals of their artificial creatures and who is responsible in the case of an accident. Due to their lack of the prerequisites for being held responsible it remains unquestioned that robots are not able to bear responsibility for anything. In paragraph 3.2 I will outline an approach for such contexts of reduced or lack of responsibility competences (ability to communicate, autonomy, judgement; cf. paragraph 1).

Within the research field of robots as moral agents, philosophers are interested in the question whether robots are themselves able to morally act and with which competences they hence have to be equipped with. Some thinkers concentrate on freedom and autonomy as prerequisite for moral agency, others on cognitive capacities, and again others on empathy and emotions.

Both fields of research within robot ethics rest on the question what morality is. Wendell Wallach and Colin Allen claim in their book *Moral Machines. Teaching Robots Right from Wrong* (2009) those beings capable of moral acting that de facto are in situations that require moral judgment. They refer to Philippa Foots famous thought experiment of the so called trolley cases [Foo67] in order to show that since the 1960s "'driverless' train systems" [Wal09, p. 14] in London, Paris, and Copenhagen morally 'judge' when they are programmed to stop whenever there are people on the tracks, even though passengers might get injured due to the abrupt halt. Of course the autonomous train, programmed with a specific algorithmic structure, is not genuinely able to act morally. However, this situation phenomenologically is comparable to those that humans might experience. That—according to Wallach and Allen—might be enough to interpret artificial systems as quasi-agents without claiming them to genuinely be moral agents in the same way than humans. I will elaborate on their approach as a version of the weak AI thesis in paragraph 3.1.

3   Ascribing Responsibility in Man-Machine-Interaction

Often people deny the possibility to ascribe responsibility to artificial systems due to their supposed lack of the necessary competences that they claim only human beings to be equipped with: Robots, following their argument, don't have the ability to communicate, autonomy, judgement, or any other morally relevant capacity. Wallach and Allen outline in *Moral Machines* (2009) an approach of functional equivalence to bypass this problem of lacking competences in artificial systems. The following two paragraphs elaborate on the role and function of responsibility within the two fields of research in robot ethics, robots as moral agents (3.1) and robots as moral patients (3.2).

3.1   *Robots as Moral Agents—Wallach's and Allen's Approach of Functional Equivalence*

In asking whether robots are to be interpreted as "artificial moral agents (AMAs)" [Wal09, p. 4], Wallach and Allen define moral agency as a gradual concept with two conditions: "autonomy and sensitivity to values" (ibid.: 25). Human beings are the genuine moral agents, but some artificial systems—e.g., an autopilot, or the artificial system Kismet—might be considered as "operational" moral agents. They are more autonomous and sensitive to morally relevant facts than non-mechanical tools, such as a hammer. However, they are still "totally within the control of [the] tool's designers and users" (ibid.: 26) and in this sense "direct extensions of their designers' values" (ibid.: 30). Only very few robots already have the status of "functional" moral agency, such as the medical ethics expert system MedEthEx. Wallach and Allen define functional morality in the sense that functional moral machines "themselves have the capacity for assessing and responding to moral challenges" [Wal09, p. 9]. They claim that "[j]ust as a computer system can represent emotions without having emotions, computer systems may be capable of functioning as if they understand the meaning of symbols without actually having what one would consider to be human understanding" (ibid.: 69).

With this notion of functional equivalence, Wallach and Allen subscribe to a version of the weak AI thesis [Sea80] that seeks to simulate certain competences and abilities in artificial systems rather than to construct robots that genuinely are intelligent, conscious, and autonomous equal to humans (that is the strong AI thesis, mistakenly ascribed to Turing 1950). According to Wallach and Allen, a strong AI understanding of autonomy is not a necessary condition for constructing AMAs. Instead they focus on the attribution of functional equivalent conditions and behavior. Functional equivalence means that specific phenomena are treated "as if" they correspond to cognitive, emotional, or other attributed competences and abilities. The question of whether artificial systems can become intelligent, conscious, or autonomous in the strong AI sense is replaced by the question to what extent the displayed competences correspond to the function they play within the moral evaluation, in this case the concept of responsibility. However, although Wallach and Allen claim the boundary between functional morality and full moral agency to be gradual with respect to certain types of autonomy, for the foreseeable future it is hard to fathom how an artificial system might achieve a functional equivalent to the genuinely human ability to set "second-order volitions" [Fra71, p. 10] for oneself and to act as "self-authenticating sources of valid claims" [Raw01, p. 23], or to be able to reflect on its own moral premises.

On a computational level these forms of autonomy and the ability to autonomously change internal states might be described by distinguishing three different types of algorithmic schemes [Loh17b]. While *determined algorithms* simply give the same output, given a particular input (independent from the sequence of states that they pass through), *deterministic algorithms* give the same output, given a particular input, in passing through the same sequences of states, while *non-determined algorithms* even have a limited variety of outputs, given a particular input, by being able to pass through different sequences of states. Potentially, machines that predominantly function based on deterministic algorithms might be located in the not-functional and not-operational sphere. They are almost closer to the non-mechanical tools such as a hammer than to the operational realm. The operational sphere might then be reached with artificial systems that predominantly function on the basis of determined (but non-deterministic) algorithms. Finally, those few robots that are predominantly structured by non-determined (and thereby non-deterministic) algorithms are to be located in the functional realm.

Let us consider three examples: Wallach and Allen define the artificial system Kismet as an operational AMA (see above). Supplementing their approach with my understanding of responsibility and the necessary prerequisites to ascribe the ability to act responsibly (cf. paragraph 1; the ability to communicate, autonomy, and judgment), Kismet possesses a rudimentary ability to communicate since it can babble in simple noises. Judgment (if one is willing to call Kismet's behavior reasonable at all) is barely recognizable in its reaction to very simple questions. The biggest challenge in regarding Kismet as an operational responsible robot is clearly its autonomy, since the relevant sub-capacities (knowledge, historicity, personhood, and scope of influence) are very limited. In its rudimentary mobility, Kismet can autonomously move its ears, eyes, lips, and head and responds to external stimuli such as voice. To conclude, Kismet is, as Wallach and Allen suggest, still completely in the operators and users control; it does not artificially learn. To call Kismet responsible might appear comparable to calling an infant or some animals responsible. However, in contrast to the rain wetting the street, Kismet might (similar to infants and animals) open a room for debate on ascribing responsibility to robots like it, although this room for debate appears to be understandably small.

Cog is a robot that can interact with its surroundings due to its embodiment. It might pass as an example for a weak functional responsible agent, since its ability to communicate as well as judgment has been greatly improved compared to Kismet and Cog's overall autonomy has evolved, since it includes an "unsupervised learning algorithm" [Bro99, p. 70]. For instance, after running through numerous trial-and-error-attempts to propel a toy car forward by gently pushing it, Cog eventually will push the car only from the front or from behind, not from the side, since it will only move then. Cog has not been programmed to solve the task in this manner but learns from experience. Due to its limited capacity to learn, one might understand it as a weak functional agent. Calling Cog responsible might be comparable to ascribing responsibility to a very young child.

Autonomous driving systems might be identified as operational rather than functional artificial agents. Whereas their communicative and judgment skills are as developed as Cog's capabilities or even further, their overall autonomy is still kept within tight limits due to their lack of learning and non-determined (non-deterministic) algorithms. Reaching this first conclusion, responsibility regarding to autonomous driving systems then has to be distributed through a responsibility network and cannot primarily be ascribed to the artificial systems themselves as I will explain in the next paragraph.

To sum up, ascribing responsibility to artificial systems is possible so far only in very restricted terms. Evolutionary learning systems are most promising; machine learning is being studied here equivalently to children learning. So, far machine learning is not possible in moral contexts (or only in weak moral contexts). Until today one cannot ascribe responsibility to artificial systems.

3.2   Robots as Moral Patients—Responsibility Networks

My conclusion of the reflections in paragraph 3.1 is that—against the backdrop of Wallach's and Allen's approach of functional equivalence—so far artificial systems are not to be seen as responsible agents since they lack the necessary competences for ascribing responsibility (the ability to communicate, autonomy, judgment), possess them only weakly in a functional sense, or even only in an operational way. To remind the reader of the etymological minimal definition of responsibility (as the normative ability to answer, based on a psycho-motivational constitution of the responsible subject in question), our traditional understanding of responsibility is clearly individualistic insofar as we always need to define a subject or bearer of responsibility (cf. paragraph 1), whether this subject is an individual or a collective. If the prerequisites are not given it is not (or only metaphorically) possible to ascribe responsibility—as to plants, children, humans with specific disabilities, or machines.

We recently find ourselves in situations in which the involved parties are not (fully) equipped with the necessary competences for bearing responsibility while we are still certain that we need to ascribe responsibility to someone. Consider again, for instance, the case of autonomous driving systems (cf. paragraph 3.1) as operational responsible agents (equivalently to the responsibility of an infant, animal, or very young child). The autonomous car might be a moral patient insofar as it is part of our moral universe and (instrumentally) morally worth considering. However, it is not a moral agent in a significant (i.e., at least functional) way. For contexts such as these I'd like to adopt Christian Neuhäuser's concept of responsibility networks [Neu14] and elaborate on it [Loh17c]. The first premise of this approach claims that every party involved in the situation is to be held responsible to the extent that it possesses the necessary prerequisites for responsibility.

Responsibility networks have the following characteristics: (a) They commonly take on an unusual scale. Due to their enormous scale it is very hard (b) to define one or more responsible subjects and other relative elements such as the normative criteria or the authority. Therefore, responsibility networks are to be found in contexts in which (c) it is unclear whether we might be able to define concrete responsibilities at all. A responsibility network (d) combines several different responsibilities and takes into account as well that (e) relational elements often overlap; consider for instance the parents' responsibility for their children (although this is not an example for a responsibility network but rather shows how relata in some cases overlap): Here, the children and their well-being are object and addressee of this responsibility [Som14, pp. 117–118]. Conclusively, the involved parties in a responsibility network usually (f) serve different positions within several responsibilities. Examples for responsibility networks are "climate responsibility" [Som14, Chap. 13], "responsibility in the global financial market system", and "responsibility in road traffic".

Within the responsibility network "responsibility in road traffic" numerous potentially responsible parties are involved dependent on the extent of the necessary competences for ascribing responsibility, such as the human drivers, the owners of the (autonomous) cars, the companies that sell (autonomous) cars, the programmers, the designers, but also the public of a society that 'decides' over a common sense of moral norms, lawyers, driving instructors, pedestrians, and eventually every in the traffic involved party. Regarding the object of the responsibility in road traffic, it is not possible to ascribe responsibility to one or a small number of subjects for "the" road traffic as a whole, since this object is too "huge" and complex for one or few persons to be fully responsible for it alone. However, we can divide several spheres of responsible acting within the responsibility network "responsibility in road traffic"—structured by different sets of norms, such as moral, legal, and political norms that define equivalent responsibility types. For all of these responsibility areas "the" road traffic serves as the overall object of responsibility but is necessarily differentiated in "smaller" and less complex objects of responsibility that different parties are answerable for in different ways. Responsibility for "the" road traffic might for instance refer to the economic and moral responsibility for getting safe, efficient, and as quick as possible from A to B, to the aesthetic responsibility for an aesthetically pleasing design of roads and sidewalks, or to the moral responsibility for preparing the children and young drivers for the moral challenges that are to be met in participating in road traffic. Within these and further responsibility constellations as part of the overall responsibility network "responsibility for road traffic" numerous authorities, addressees, and normative criteria are to be defined.

Currently, an autonomous driving system that is to be identified only as a very weak artificial responsible agent (as an artificial operational agent) cannot fill the subject position of a responsibility within the responsibility network "responsibility for road traffic" due to several more qualified potential (human) subjects of responsibility. However, such an artificial system could be identified as object or even addressee of one or more responsibilities and via this be included within this responsibility network. To conclude, in this manner it is possible to integrate robots as moral patients in responsibility constellations—even in challenging situations that require the complex structure of a responsibility network.

References

[Bro99]   Brooks, R. A., Breazeal, C., Marjanović, M., Scasselatti, B., & Williamson M. M. (1999). The cog project, building a humanoid robot. In C. Nehaniv (Ed.), *Computation for metaphors. Analogy, and agents* (pp. 52–87). Springer, Wiesbaden.

[Flo04]   Floridi, L. & Sanders, J. W. (2004). On the morality of artificial agents. *Minds and Machines, 14*, 349–379.

[Foo67]   Foot, P. (1967). Moral beliefs. In P. Foot (Ed.), *Theories of ethics* (pp. 83–100). Oxford University Press.

[Fra71]   Frankfurt, H. (1971). Freedom of the will and the concept of a person. *Journal of Philosophy, 68/1*, 5–20.

[Loh17a]   Loh, J. (2017). Strukturen und Relata der Verantwortung. In Heidbrink, L., Langbehn, C., & Loh, J. (Eds.), *Handbuch Verantwortung* (pp. 35–56). Springer VS, Wiesbaden.

[Loh17b]   Loh, J. (2017). Roboterethik. *Information Philosophie, 1*, 20–33.

[Loh17c]   Loh, J., & Loh, W. (2017). Autonomy and responsibility in hybrid systems. The example of autonomous cars. In P. Lin, K. Abney, & R. Jenkins (Eds.), *Robot ethics 2.0. From autonomous cars to artificial intelligence* (pp. 35–50). Oxford University Press.

[Neu14]   Neuhäuser, C. (2014). Roboter und moralische Verantwortung. In E. Hilgendorf (Ed.), *Robotik im Kontext von Recht und Moral* (pp. 269–286). Nomos, Baden-Baden.

[Raw01]   Rawls, J. (2001). *Justice as fairness. A restatement*. Harvard University Press.

[Sea80]   Searle, J. R. (1980). Minds, brains and programs. *Behavioral and Brain Sciences, 3/3*, 417–157.

[Som14]   Sombetzki, J. (2014). *Verantwortung als Begriff, Fähigkeit, Aufgabe. Eine Drei-Ebenen-Analyse*. Springer VS, Wiesbaden.

[Wal09]   Wallach, W. & Allen, C. (2009). *Moral machines. Teaching robots right from wrong*. MIT Press, Cambridge, Massachusetts, London.

Psychological Issues for Developing Systems for Older Users

Rebecca Wiczorek

Abstract When designing technology for older users, several psychological aspects need to be taken into account. With increasing age, certain abilities decline making interaction with interfaces more challenging and time consuming. Understanding problems older users face when using new technologies is the key to improving the design according to their requirements. Several rules can be applied and techniques can be used to support perception and analysis of information through the interface as well as the appropriate action selection and action implementation. This chapter gives a short overview of psychological factors relevant for successful use of technical devices and presents some examples how to support older users by compensating for their age-related declines.

1 Introduction

When developing technology for older adults, two main areas can be distinguished. First, there is the (re-)design of already existing devices, such as computers, smartphones, etc. for the specific target group of older users considering their special needs. The second type of application would be the development of special devices tailored to the age-related needs of the older population. The main difference between the two approaches is that in the case of re-design, the focus mainly lies on the interface, while the functionality remains the same. When developing new systems for the older target group, system functionalities have to be defined, developed, programmed, and tested. The current chapter focusses on the (re-)design of interfaces as the relevant part of interaction. Interfaces can be divided in input devices, such as a computer mouse or output devices such as loudspeakers or a screen. A lot of interfaces combine the two types via touchscreen.

When talking about interface design for older people, two different approaches can be applied. The planned interface should either be exclusively used by older people

R. Wiczorek (✉)

Department of Psychology and Ergonomics, Technical University of Berlin, Berlin, Germany
e-mail: wiczorek@tu-berlin.de

© Springer Nature Switzerland AG 2018   109
A. Karafillidis and R. Weidner (eds.), *Developing Support Technologies*, Biosystems &
Biorobotics 23, https://doi.org/10.1007/978-3-030-01836-8_10

or it should be designed in a way to be used by people of all age groups. The so-called "design for all" is required for several applications that are used by different people and must therefore fulfill the needs of everyone. Examples are cash machines, ticket machines, and websites. For other devices, however, "design for all" might not be the best solution because what increases usability for older people (e.g., extra sized symbols) may decrease the usability for younger people (e.g., unnecessary need for scrolling). One feasible solution can be the adaptability to customer specification.

2 Older Adults' Psychological Characteristics Relevant for Technical System Use and Design

Designing devices to fit the needs of older users is challenging because the group of older users is very heterogeneous. Some experience reductions of their abilities already in early old age (e.g., 65 and younger), whereas others remain vital and keep a high performance-level until they are very old (85 and older). Furthermore, the decline of a certain aspect does not necessarily imply the decline of others. Not all cognitive functionalities are strongly interrelated. However, aging is a universal progressive process experienced by every human being. Thus, every older person will face some or all of the presented problems at some point in life.

Whenever interacting with the environment, with other people, or with technical agents, humans pass through the same stages of information processing (see Fig. 1). Those stages are: perception of information, analysis of information, decision-making/response selection, and action implementation (e.g., [Wic15]).

With increasing age, performance in all the four stages decreases. Sensory perception diminishes, especially of eyes and ears, understanding new situations becomes more challenging, comparison of alternative options is slower and less accurate, and execution of movements becomes less precise and sometimes painful. Thus, the interaction with technical systems through input and output devices is more challenging for older adults.

Understanding the deficits of older people can help identifying difficulties they face in interaction with technical interfaces. When knowing the underlying mechanisms involved, it is possible to (re-)design interfaces that support older users by compensating for their age-related declines. In the following, a short overview of psychological factors of older people regarding perception, cognition, and motoric aspects relevant for successful use of technical devices is given. The chapter is mainly

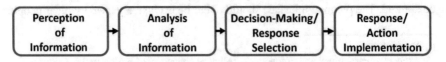

Fig. 1. Four stages of human information processing in dependence on [Wic15]

based on the work of several authors presenting summaries of psychological aspects of aging [Bir06, Cra08, Stu12] and handbooks for the design for older adults [Fis09, Pak11].

3 Perception of Information and Sensory Abilities

All five sensory systems decline with age. The three most relevant for interaction with technical devices are the visual, the auditory, and the haptic channel. *The perception can be increased by using multi-modal stimulation.*

It is not only the acuity, but several functions involved in *vision* that decrease with age, as for example contrast sensitivity, dark adaptation, color perception (due to yellowing), etc. *Make sure to provide sufficient light condition, size, contrast, and use colors not affected by changes in yellow for improving perception of information.*

The most problematic change in *hearing* is the loss of sensitivity for high frequencies affecting perception of speech and several sounds. *To increase comprehension, signal-to-noise ratio should be increased, frequencies above 500 Hz should be avoided and natural should be preferred over synthetic sounds.*

4 Analysis of Information and Cognitive Abilities

In order to interact with an interface, it is not only necessary to see (or hear) the content, but also to understand its meaning. The first important cognitive ability needed is *attention*. In order to properly begin analyzing information, selective attention must be given to the item or items of interest. Therefore, distracting information has to be inhibited, a mechanism that works poorer in older people due to changes of the frontal lobe. In addition, they have more difficulties when being confronted with multi-tasking demands requiring diverted attention. *Thus, it is important to eliminate anything not relevant to the task and to avoid parallel tasks.*

Furthermore, the useful field of view (UFOV) declines with age (see [Bal90]). That is the part of the visual field where the attentional focus works. The smaller the UFOV, the fewer items can be processed in parallel. Older people are more likely to make a serial visual search, paying attention to one (or only a few) item(s) after the next taking more time. *In order to reduce long visual search time, avoid clutter by only displaying relevant information.*

The next important cognitive component is the *working memory*, which is a limited short-term storage unit for a small number of items that must be manipulated (e.g., mathematical operations) and/or remembered. When aging, older users face the two problems of reduced capacity and increased processing time. While younger people are able to process and remember seven ± two items, older adults process and remember four or less items. In addition, the general-slowing theory [Sal96] states that the speed with which each signal is being processed is reduced in older adults.

Thus, the 'age x complexity' hypothesis explains why increasing the number of items needed for a certain operation in working memory proportionally increases the processing time. *In order to avoid errors of working memory, the number of items that must be handled or remembered should be as low as possible.*

5 Decision-Making/Response Selection and Cognitive Abilities

The accurate analysis of what has been perceived is crucial to understand what is presented by an interface. However, for the successful interaction users must be able to comprehend the information displayed. In order to do so, they need *reasoning abil- ity*, the competence of deductive conclusion of the available options by interpreting the current situation. Like most of the other cognitive abilities, deductive reasoning declines with age [Sch05], but some simple techniques may serve as countermea- sures. *For improving sense-making, the most crucial thing is consistency achieved by tying only one functionality to each button (or switch, etc.) and keeping it during the different modes. Additionally, similar items should share identical properties, such as color, shape, etc. and should be grouped together.* Even though reasoning ability is usually defined with regard to sense-making in *new* situations, older adults often face the additional disadvantage of being unfamiliar with technical devices in general. Thus, it is even more challenging to understand the possible options of manipulation (e.g., using a touchscreen). *To support action-selection, the input options should possess affordance characteristics regarding size, shape, color etc.*

One more specific intellectual aspect also declining with age is the *spatial ability* needed for navigation in the environment as well as through menus of an interface. Older users are more likely to lose track when navigation becomes complex. *For avoiding older users getting confused when navigating, hierarchy should be kept flat and representations of the underlying structure such as a graphical organization of the menu should be provided if possible.*

The ability that is least affected by age and can even improve is the *verbal ability*. It is therefore a valuable source for compensating for other declines. *Use letters and text instead of symbols and pictures to improve comprehension. Add verbal explanations such as manuals or other help options (e.g., mouse-over), whenever possible to facilitate interaction.*

6 Response/Action Implementation and Motor Abilities

When interacting with input devices precise movements have to be fulfilled in order to achieve the desired goal. With age, the *accuracy* of motion declines and *timely coordination* becomes more difficult. The first causes errors with precise pointing interaction, whereas the latter can be problematic when coordinated actions such as double clicks or scrolling are required. *To support older users' manual interaction items (e.g., buttons) should be increased in size with sufficient space between them. Whenever possible, direct interaction as with touchscreens should be preferred over indirect manipulation with devices, such as a computer mouse.*

7 The Design Process

When comparing all requirements supporting older users' interaction with technical interfaces, it becomes clear that some of them are contradictory:

- The number of items that need to be scanned visually should be reduced, but at the same time the hierarchy of menus shall be flat not deep, which leads to an increase of the number of items per level.
- When a touchscreen with extra sized buttons is used as required to improve manual interaction, fewer buttons load on a page, but switching between pages increases the load of working memory.
- Distracting elements and multi-tasking demands should be avoided to support attention allocation, but the requirement of a graphical organization of the menu structure with the aim to support spatial navigation contradicts the first rule.

The examples given above point out that designing for older adults does not only consist of following simple rules. It is a process requiring careful consideration of contradictory requirements, acceptance of necessary compromises, and, most important, an individual answer for each case. It is important to find a specific solution that works for the current system, the current task, and the current target group of users.

Certain steps of the *user-centered design* approach can help to identify the optimal solution for a currently developed system:

- The first step is an analysis of the task, the users, and the environment. This can be done with a hierarchical task analysis, interviews, questionnaires, observations, etc.
- Based on the results, designers can decide which tasks or task aspects are most important (e.g., most frequently carried out, most crucial for safety, etc.).
- Accordingly, priority shall be given to design aspects supporting the most important tasks.
- Before implementing the new solution, the chosen design should be evaluated. This is usually done by a usability test. This test investigates the effectiveness, efficiency, and the satisfaction of the older users when completing a task with the new or re-designed interface/system. Requirements for the test can be derived from the analysis carried out in the beginning of the design process.

8 Example: Re-Designing a Smartphone for Older Users

The example is hypothetical and refers to a system most readers are familiar with, the smartphone. Obviously, this is a very complex system and the example is striking and simplistic, but hopefully it is nevertheless helpful by pointing to the most important aspects of user-centered design for older people, i.e., prioritization and compromises.

In a first step the *analysis* should be carried out. Interviews and questionnaires provide knowledge regarding the most important tasks as well as the least important tasks. Additionally, problems that were faced with already existing devices may help to avoid certain errors in the future product.

The hypothetical *results* show a lot of differences between older users and the conventional target group: The two tasks they are most interested in are 'making phone calls' and 'storing and watching pictures' received from their family. The tasks they carry out less often are texting, surfing in the internet, and taking pictures of themselves. The most crucial problems they have faced before are insufficient audio quality and getting lost in the menu. Additionally, older people are not willing to spend as much money for their smartphone as the average younger user.

Thus, in the hypothetical design process, we *decide* to develop an economic smartphone that mainly supports the task of making phone calls and the task of storing and watching pictures. Users shall not get lost in the menu and have easy access to relevant functionalities such as changing the brightness.

Accordingly, we give *priority* to the speakers over the camera. Additionally, we can develop a software solution to further improve the sound quality for older people (e.g., increasing high frequencies). To avoid getting lost, items representing important tasks (e.g., dialing, contacts, pictures) can be placed in the home screen, represented by oversized buttons labeled with readable text. To *compromise* about other functions, they may be placed in a second screen reachable with an 'other functions' button.

Within a usability test with a prototype, we can *evaluate* our hypothetical solution by making older participants carry out the relevant tasks with the new system in comparison to conventional smartphones.

The hypothetical product we designed is not the *perfect* smartphone that satisfies all types of users and is best for all tasks. The current phone does not have a back camera, it does not allow to place lots of applications in the home screen, it has not a very minimalistic design, etc. However, in the best case it represents the *optimal* solution for older users, because it supports the tasks important to *them* and minimizes problems of the *specific target group*.

9 Summary

Declines of physical and cognitive functions often make the interaction with technical devices more challenging for older people. When designing for this specific target group it is helpful to be familiar with psychological issues relevant for the design of technical systems as well as knowing adequate countermeasures to support the older users. However, it is not sufficient to follow simple rules as they are often contradictory. Thus, it is necessary to compromise about certain functionalities by giving priority to the tasks most important for the older user group. Methods of user-centered design can help to structure the design process and to make the relevant design decisions.

Acknowledgements This research is part of the project "FANS—Pedestrian Assistance System for Older Road Users" funded by the German Federal Ministry of Education and Research (BMBF, funding No. 16SV7112) and supervised by VDI/VDE Innovation + Technik GmbH.

References

[Bal90]   Ball, K. K., Roenker, D. L., & Bruni, J. R. (1990). Developmental changes in attention and visual search throughout adulthood. *Advances in Psychology, 69*, 489–508.

[Bir06]   Birren, J. E., & Schaie, K. W. (2006). *Handbook of the psychology of aging.* Amsterdam: Elsevier Academic Press.

[Cra08]   Craik, F. I., & Salthouse, T. A .(2008). *The handbook of aging and cognition.* New York: Psychology Press.

[Fis09]   Fisk, A. D., Czaja, S. J., Rogers, W. A., Charness, N., & Sharit, J. (2009). *Designing for older adults: Principles and creative human factors approaches.* New York: CRC press.

[Pak11]   Pak, R., & McLaughlin, A. (2011). *Designing displays for older adults.* New York: CRC press.

[Sal96]   Salthouse, T. A. (1996). The processing-speed theory of adult age differences in cognition. *Psychological Review, 103*, 403.

[Sch05]   Schaie, K. W. (2005). What can we learn from longitudinal studies of adult development? *Research in Human Development, 2*, 133–158.

[Stu12]   Stuart-Hamilton, I. (2012). *The psychology of ageing: An introduction.* London: J. Kingsley Publishers.

[Wic15]   Wickens, C. D., Hollands, J. G., Banbury, S., & Parasuraman, R. (2015). *Engineering psychology and human performance.* Psychology Press.

Attention Models for Motor Coordination and Resulting Interface Design

Bettina Wollesen, Laura L. Bischoff, Johannes Rönnfeldt and Klaus Mattes

Abstract In the Industry 4.0 interface designs need to be adjusted to cognitive and sensorimotor abilities of humans in order to ensure a faultless and ergonomic human-machine interaction. The perception and processing of stimuli as well as the reactive motor planning and response of humans is essential for the exchange of information. Attention processes play an eminent role in both the processing of stimuli and the motor response. This chapter presents the current state of research regarding attention models for perception and motor control. Various attention theories agree that the execution of motor-cognitive tasks depends on the task setting and the task conditions (e.g., the complexity or sensory modality conditions). These findings should be actively integrated into design processes of interfaces for human-machine interaction to avoid negative consequences.

1 Introduction

Employees who work in the field of human-machine interaction need to coordinate their cognitive and sensorimotor abilities to make correct decisions during working processes (e.g., the monitored inspection of components produced on assembly lines). Every action during a working process needs the coordination of motor skills including the related components of sensory input and transmission, cognitive control and motor planning, control and performance.

All sensations from our everyday surroundings—whether it is a sound that signals the arrival of a new e-mail or a visual sensation needed to recognize and sort out defective components produced on assembly lines—they all are perceived via several receptors (visual, auditory and kinesthetic system). For processing these information, reactive motor actions have to be taken, which must be regulated and implemented (e.g., clicking on an e-mail or grabbing a defective component of an assembly line) [Ros08].

B. Wollesen (✉) · L. L. Bischoff · J. Rönnfeldt · K. Mattes
Institute of Human Movement Science, University of Hamburg, Hamburg, Germany
e-mail: bettina.wollesen@uni-hamburg.de

© Springer Nature Switzerland AG 2018
A. Karafillidis and R. Weidner (eds.), *Developing Support Technologies*, Biosystems & Biorobotics 23, https://doi.org/10.1007/978-3-030-01836-8_11

117

However, not all reactions to a specific stimulus automatically lead to a specific sequence of actions. It is more likely that after a stimulus is sensed, specific movements have to be coordinated and deliberately implemented; hence motor-cognitive control and movement requires attention and particular interaction processes.

2 Organization of Interaction Processes of Perception and Motor Control and the Role of Attention

At the modern workplace, employees are continuously faced with new tasks and novelty. Willed motor action and the execution of plans require a *supervisory attentional system* (SAS) [Nor86]. For every initiated movement, there has to be differentiation between automatic and controlled processes.

Automatic movement control is based on patterns that are stored in the long-term memory and are activated automatically by specific impulses and initiate routine actions. Automatic processing of those patterns is fast, demands few resources and can be done subconsciously. The lateral inhibition prevents movement patterns to run simultaneous, and even prioritizes established patterns that have proved to be adequate motoric reactions to certain stimuli [Nor86].

In contrast to automatic movement patterns, the SAS influences controlled movements: It adjusts the movement patterns to primary objectives, activates matching movements and inhibits those that do not fit. Hence, new learned movement patterns and those which have to be adjusted require special attention [Kar12]. The SAS must inhibit automatic patterns, especially in situations, in which automatic movements have to be inhibited (this can be the case e.g., when ambulance service drivers must instead of stopping, run red lights).

Other models as the *frontal-lobe executive model* (FLE [Dun86]) and the *strategic response-deferment* (SRD [Mey97]) are still discussed diversely. Supporters of the FLE assume that three sub-components influence the executive control of actions (orientation on the primary target, analyzing processes of expected meaning and action and the related structure to activate movements). On the other hand, scientists who support the SRD expect that responses to the task-stimuli and the organization of the tasks are integrated into the cognitive operations according to the order they will be processed. Secondary tasks stay in the working memory and are not completed until prioritized tasks are.

In their review, Wulf et al. [Wul10] describe the influence attention has on performance of movements in respect of effectiveness (precision and consistency) and efficiency (muscle activity, physical effort, cardiovascular stress). Results are far better when the attention is focused on external effects of the motion execution than when the execution of movements is focused internally.

The lack of attention, even divided attention due to so-called multi-tasking can lead to an inadequate choice of, and even to an incorrect execution of movement sequences, because task organization requires different cognitive processes. Yet the capacity of the working memory is limited [Spi08].

3 Problems Caused by Multi-tasking

During the workday, employees frequently switch between activities such as making phone calls, using computers, reading, or supervising machine operations [Spi08]. That is why, for many jobs multi-tasking is required, when executing parallel or serial tasks. Multi-tasking can be classified in four categories [Sal05]:

1. Serial discrete tasks (task-switching).
2. Simultaneous discrete tasks (overlapping tasks).
3. Continuous tasks (interrupted only by occasional discrete tasks).
4. Composite (simultaneous) continuous tasks.

This classification shows that in the working place, there is a close connection between multi-tasking and interruptions [Jan15]. If two or more tasks are to be completed at the same time, the needed time for the completion can be expected to rise significantly [Pas00, Rub01]. The interruption of tasks is also an uncontrollable stressor, which requires additional psychological resources in order to manage the completion of tasks [Spi08].

Alongside interruptions, the complexity of secondary tasks plays an important part. Colcombe and Kramer [Col03] differentiate between four task classes. Starting with the least amount of cognitive effort needed and ending with the most, the four classes are: (1) processing speed, (2) visuospatial, (3) controlled processing, and (4) executive control.

The quality of the work can be expected to decline significantly: the more tasks have to be completed simultaneously and the longer a multi-tasking situation lasts, the higher the loss of quality gets [Jan15, Rub01].

Various theoretical models describe the mechanisms of motor-cognitive interactions and susceptibility in situations of complex actions. They all underlie the idea that dual-task compete for attentional resources, because cognitive processes run serially and not simultaneously.

The theory of the so-called *central bottleneck* hypothesizes that cognitive processes run serially, without any exception. It implies that the processing of information from different task demands is limited (bottleneck), because not all information can be processed in the working memory simultaneously. The processing of the second task can only begin, when the first task has been processed already. That is why the processing time for dual-task is longer than for single-tasks [Pas00].

However, the *attentional resource theory* indicates that motor-cognitive interferences occur in multi-tasking situations, because the subtasks compete for attentional resources, which are limited [Kah73]. Yet, it is striking that, when executed simultaneously, controlled performance patterns show a higher loss of quality than automatic ones. The loss of quality even rises, if the tasks that are to be performed resemble each other.

These observations helped to supersede unspecific theories of resources by concepts like Wickens' *4-dimensional multiple resource model*. Among others, Wickens' model shows that the interference rises, whenever tasks require similar sensory modalities and information channels [Wic80].

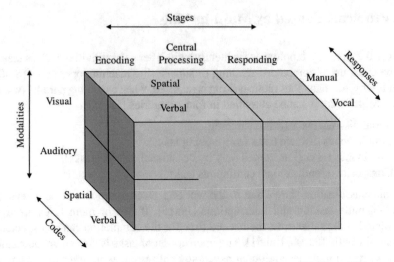

Fig. 1 Wicken's 4-dimensional multiple resource model adapted from [Wic02, p. 163]

This model also suggests that the cognitive system has limited capacities. The entire capacity is made up from different, independent singular capacities. The model is to be understood as a cube with multiple dimensions (see Fig. 1). This means that perception modalities, such as acoustic or visual ones, are differentiated and hence the reaction to those modalities can be either manual or vocal.

Wickens postulates that less interferences occur, when different modalities or reactions are taking place in parallel contrary to cases, when tasks require similar modalities or reactions. This means that it is easier for the human brain to process one acoustic and one visual signal than to process two acoustic signals [Wic80].

Therefore, an organization process is needed in the central nervous system to distribute attention resources. It is assumed that such processes are run by the mentioned SAS [Nor86] or that they are executed via a central source [Bad96]. This assures that the resources can be strategically distributed according to (a) the availability of resources, (b) the priority of the task and (c) according to the requirements for the task. The organization process is subdivided into 'task switching', 'memory updating' and 'response inhibition' [Bad96, Miy00, Str14]. Multitasking has also been suggested to be the executive function itself (executive function 'dual tasking') rather than being a process made up from the three functions named above [Str14, Enr13].

In multisensory research, there is now an ongoing debate whether there are independent singular attentional capacities for each sensory modality (as postulated by Wickens) or whether attentional resources are shared across sensory modalities. A recent review conducted by Wahn and König [Wah17], suggests that for the visual and auditory sensory modalities, distinct resources are recruited when humans perform object-based attention tasks, whereas for the visual and tactile sensory modalities, partially shared resources are recruited. Attentional resource allocation across sensory modalities is therefore task-dependent.

With advancing age, routine actions in everyday life, such as walking, driving, or typing on a computer's keyboard require more cognitive control and attentional resources [Yog12, Woo02].

Research has shown a correlation between age and poor results when solving even very simple visual reaction time tasks. The results decrease in quality with increasing age of test persons and with rising difficulty of the tasks, especially when the solution requires executive functions [Foz94]. Hence, demanding visual tasks have a negative impact on a person's motor skills [Beu12].

Apart from research about multitasking situations during locomotion and postural control, multiple papers have also explored performance decrements of everyday skills (e.g., driving). Age-related performance decrements during the multitasking situation of driving a car were detected especially when the drivers had to complete visual tracking tasks. When given mathematic tasks, drivers slowed down and overlooked street road signs [Cha05]. Talking on the phone also impacted the speed while driving and the ability to change the speed. Additionally, it leads to deviations of steering motions [Shi05]. Körber', Gold, Lechner and Bengle [Kör16] conducted research to investigate whether elderly drivers reacted differently to obstacles, when they were confronted with a high volume of traffic or verbal multitasking. Although the time elderly people needed for a reaction was comparable to the reaction times of young drivers, it was evident that elderly drivers decelerated more rapidly and more often. The precision of movements for bimanual coordination requires executive control and therefore relies on the allocation of attention resources [Ban10].

Additionally, findings concerning motor-cognitive interaction processes of the fine motor skills of elderly people should be taken into consideration, when designing interfaces.

Hand-Eye Coordination

Hand-eye coordination is often disturbed because the strategy for grabbing an object needs to be altered: the targeted object has to be visually focused for a longer period of time, if movements are to be executed extremely precise [Coa16].

Grabbing and Pointing Movements

Precision of grabbing and pointing movements can cause movements to be executed slower. They also have a higher movement variability [Yan98]. Especially pointing movements become less precise and disharmonious [Sei02].

Bimanual Coordination, Grabbing and Manipulating

Asynchronous movement of the hands can cause deficits in grabbing objects bimanually and in continuous, cyclic movements; but it also results in generally prolonged execution of movements and force control [Wis00, Vie15, Gil03, Par12, Ola07].

When reviewing the presented models, it can be concluded that managing motor-cognitive tasks in everyday life depends on the task setting and on the task requirements (e.g., the complexity or sensory modality conditions). This knowledge should be directly implemented in designing processes.

4 Mistakes and Situation Awareness

Endsley's concept [End95] has become the most used models to identify possible mistakes made by users, who process information with the help of computer-based systems. His model of *situation awareness* identifies three consecutive aspects of a cognitive process:

- Perceiving all relevant objects in a given environment (e.g., identifying objects indicated on the monitor).
- Understanding the meaning of all identified objects and the interpretation of the situation (e.g., users have to understand displayed information and maybe even need to integrate information into their existing knowledge).
- Predicting the changes of the object's state and of the environment for a certain time span.

Often, cognitive processes result in decisions, which may lead to reactive actions. For the actions of users, it is of utmost importance that they are able to anticipate future conditions of the system. Mistakes often result from conscious or unconscious anticipatory processes. However, they can occur in all the three cognitive processes [End95].

To avoid mistakes, the cognitive abilities of humans should be taken into consideration when designing interfaces. An intuitive design can reduce the cognitive costs that are needed to pass through consecutive processes [End95].

In the following the consequences that result from the presented attention models and from the interaction between attention and movement planning and movement execution for human-machine interaction, will be presented.

5 Consequences for the Design of Interfaces in Human-Machine Interactions

The primary objective of an ergonomic design for human-machine interaction should be the adjustment of technological possibilities to the skills and abilities of humans [Jai07, Sch10]. For example, display systems should be constructed in a way that allows information to be perceived, processed, and interpreted with as little mental resources as possible.

As the SAS [Nor86] indicates, fewer attention resources are needed, when designs of interfaces are intuitive, which means that the design is oriented on expectations and experiences that are stored in the long-term memory and hence activate automatic action processes. Interfaces with non-intuitive designs require the activation of supervising attention systems to inhibit automatic action processes. Therefore, they have higher attention costs and are not ergonomic [Gor14].

Additionally, the described limitations of human's capacity to act when confronted with multitasking, which result from the necessary allocation of limited attention, is to be taken into consideration when designing interfaces. As the theoretical discussion of Wickens' *multiple resource model* [Wic80] has shown, there is a high demand for multimodal intersections between humans and machines.

The main characteristic of unimodal intersections is that they have one specific channel to send information (mostly visual information, such as the digital dashboard of cars or the screen of a computer) and one channel to receive information (mostly a manual one, e.g., the steering wheel or the keyboard). Unimodal intersections are sufficient for simple tasks. In case of complex tasks, multimodal intersections reduce dual task cost and therefore lower potential risks of making mistakes. By providing various displays and input elements, a multimodal system offers the possibility to handle machines ergonomically, even during multitasking conditions [Gär00, Tro07]. Schlick et al. [Sch10], who follow Wicken's model, assume that a perceptive-cognitive task, which requires the user to receive spatially coded information via auditory channels and transmit them verbally, can be easily combined with a reactive action, which requires the user to take in auditory coded information visually and transmit them manually. Moreover, Schlick et al. believe that tasks, which access different modalities have better chances to be combined than tasks that access similar modalities [Sch10].

Research has proved the advantages of multimodal interfaces: they prevent mistakes, they help users to correct mistakes, and they extend the range of human-machine communication in a way that makes alternatives for communication in various situations and environments available [Ovi00].

Also, elderly people, who show higher interferences in multitasking situations would profit from a wider range of communication channels and a modest use of mental resources achieved by multimodal systems. In special regard to elderly employees, information channels should be added to tasks that are visually challenging, because—as research suggests—when elderly people use visual channels in multitasking situations, interferences are significantly higher than in cases, where other channels that process information are available [Beu12]. For elderly employees, it is also to be taken into consideration that the precision of movements and their reaction time changes considerably for manual tasks. Here trajectories could assist the movements. This is a solution that younger employees would also profit from, if the tasks lead to muscle fatigue.

In the past decade, researchers in the field of human-machine interaction have discussed the possibility to relieve the visual channel that is used most frequently in the perception of information [Jai07, Sch10, Gor14]. However, the channel is crucially needed to keep postural control while executing tasks when standing or in motion. Since acoustic stimuli—especially in an industrial environment with high noise pollution—are perceived later than visual impulses and additionally are often explicitly interconnected with warnings, it should be considered to use tactile or haptic feedback as a substitute for the visual channel. In that way, the visual would be released and multitasking cost could be reduced.

6 Summary and Perspectives

When two or more tasks are executed simultaneously, the time needed to complete the tasks increases considerably while the quality of the performed tasks decreases significantly. The decline in performance is evident for cognitive and motor skills.

Multiple studies have proved that multitasking conditions impact the processing of information and the motoric responsiveness. The losses during multitasking situations result from a necessary allocation of a limited capacity of attention.

Various attention theories offer different explanations. However, they all agree that the execution of motor-cognitive tasks depends on the task setting and the task conditions (e.g., complexity, stimulus-response conditions). For example, Wickens assumes that tasks, which access differing modalities of attention and reaction can be combined easier than tasks for which similar modalities are used.

These findings should be actively integrated into design processes of interfaces for human-machine interaction to avoid negative consequences. The reviewed results about the design of ergonomic user interfaces and work places also play an eminent role for the industry 4.0. In the future, work places will be dominated by an everyday interaction between humans and machines. In order to preserve the efficiency and health of employees and to prevent mistakes, an adjustment of attention capacities and cognitive processes is of utmost importance.

References

[Bad96]   Baddeley, A. (1996). Exploring the central executive. *The Quarterly Journal of Experimental Psychology Section A, 49*(1), 5–28.
[Ban10]   Bangert, A. S., Reuter-Lorenz, P. A., Walsh, C. M., Schachter, A. B., & Seidler, R. D. (2010). Bimanual coordination and aging: Neurobehavioral implications. *Neuropsychologia, 48*(4), 1165–1170.
[Beu12]   Beurskens, R., & Bock, O. (2012). Age-related deficits of dual-task walking: A review. *Neural Plasticity* 131608.
[Cha05]   Chaparro, A., Wood, J. M., & Carberry, T. (2005). Effects of age and auditory and visual dual tasks on closed-road driving performance. *Optometry and Vision Science: Official Publication of the American Academy of Optometry, 82*(8), 747–754.
[Coa16]   Coats, R. O., Fath, A. J., Astill, S. L., & Wann, J. P. (2016). Eye and hand movement strategies in older adults during a complex reaching task. *Experimental Brain Research, 234*(2), 533–547.
[Col03]   Colcombe, S., & Kramer, A. F. (2003). Fitness effects on the cognitive function of older adults: A meta-analytic study. *Psychological Science, 14*(2), 125–130.
[Dun86]   Duncan, J. (1986). Disorganisation of behaviour after frontal lobe damage. *Cognitive Neuropsychology, 3*(3), 271–290.

[End95] Endsley, M. R. (1995). Toward a theory of situation awareness in dynamic systems. *Human Factors: The Journal of the Human Factors and Ergonomics Society, 37*(1), 32–64.

[Enr13] Enriquez-Geppert, S., Huster, R. J., & Herrmann, C. S. (2013). Boosting brain functions: Improving executive functions with behavioral training, neurostimulation, and neurofeedback. *International Journal of Psychophysiology: Official Journal of the International Organization of Psychophysiology, 88*(1), 1–16.

[Foz94] Fozard, J. L., Vercryssen, M., Reynolds, S. L., Hancock, P. A., & Quilter, R. E. (1994). Age differences and changes in reaction time: The Baltimore longitudinal study of aging. *Journal of Gerontology, 49*(4), 179–189.

[Gär00] Gärtner, K.-P. (Ed.). (2000). Multimodale Interaktion im Bereich der Fahrzeug- und Prozessführung: 42. Fachausschusssitzung Anthropotechnik der Deutschen Gesellschaft für Luft- und Raumfahrt e.V., 24. und 25. Oktober 2000, München. Fachausschusssitzung Anthropotechnik der Deutschen Gesellschaft für Luft- und Raumfahrt: Vol. 42. Bonn: DGLR.

[Gil03] Gilles, M. A., & Wing, A. M. (2003). Age-related changes in grip force and dynamics of hand movement. *Journal of Motor Behavior, 35*(1), 79–85.

[Gor14] Gorecky, D., Schmitt, M., Loskyll, M., & Zuhlke, D. (2014). Human-machine-interaction in the industry 4.0 era. In *2014 12th IEEE International Conference on Industrial Informatics (INDIN)* (pp. 289–294).

[Jan15] Janssen, C. P., Gould, S. J. J., Li, S. Y. W., Brumby, D. P., & Cox, A. L. (2015). Integrating knowledge of multitasking and interruptions across different perspectives and research methods. *International Journal of Human-Computer Studies, 79*, 1–5.

[Jai07] Jaimes, A., & Sebe, N. (2007). Multimodal human–computer interaction: A survey. *Computer Vision and Image Understanding, 108*(1–2), 116–134.

[Kah73] Kahneman, D. (1973). *Attention and effort. Prentice Hall series in experimental psychology*. Englewood Cliffs: Prentice Hall.

[Kar12] Karnath, H.-O., & Thier, P. (2012). *Kognitive Neurowissenschaften*. Berlin, Heidelberg: Springer, Berlin Heidelberg.

[Kör16] Körber, M., Gold, C., Lechner, D., & Bengler, K. (2016). The influence of age on the take-over of vehicle control in highly automated driving. *Transportation Research Part F: Traffic Psychology and Behaviour, 39*, 19–32.

[Mey97] Meyer, D. E., & Kieras, D. E. (1997). A computational theory of executive cognitive processes and multiple-task performance: Part 1. Basic mechanisms. *Psychological Review, 104*(1), 3–65.

[Miy00] Miyake, A., Friedman, N. P., Emerson, M. J., Witzki, A. H., Howerter, A., & Wager, T. D. (2000). The unity and diversity of executive functions and their contributions to complex "Frontal Lobe" tasks: A latent variable analysis. *Cognitive Psychology, 41*(1), 49–100.

[Nor86] Norman, D. A., & Shallice, T. (1986). Attention to action. In R. J. Davidson, G. E. Schwartz & D. Shapiro (Eds.), *Consciousness and self-regulation* (pp. 1–18). Boston, MA: Springer US.

[Ola07] Olafsdottir, H., Zhang, W., Zatsiorsky, V. M., & Latash, M. L. (2007). Age-related changes in multifinger synergies in accurate moment of force production tasks. *Journal of applied physiology (Bethesda, Md.: 1985), 102*(4), 1490–1501.

[Ovi00] Oviatt, S., & Cohen, P. (2000). Perceptual user interfaces: Multimodal interfaces that process what comes naturally. *Communications of the ACM, 43*(3), 45–53.

[Par12] Parikh, P. J., & Cole, K. J. (2012). Handling objects in old age: Forces and moments acting on the object. *Journal of applied physiology (Bethesda, Md.: 1985), 112*(7), 1095–1104.

[Pas00] Pashler, H. (2000). Task switching and multitask performance. In J. Driver & S. Monsell (Eds.), *Control of cognitive processes. Attention and performance XVIII* (pp. 277–305). Cambridge, MA: MIT Press.

[Ros08] Rosenbaum, D. A. (2008). *Human motor control*. San Diego, CA: Academic Press.

[Rub01]   Rubinstein, J. S., Meyer, D. E., & Evans, J. E. (2001). Executive control of cognitive processes in task switching. *Journal of Experimental Psychology: Human Perception and Performance, 27*(4), 763–797.

[Sal05]   Salvucci, D. D. (2005). A multitasking general executive for compound continuous tasks. *Cognitive Science, 29*(3), 457–492.

[Sch10]   Schlick, C. M., Bruder, R., & Luczak, H. (2010). *Arbeitswissenschaft*. Berlin, Heidelberg: Springer, Berlin Heidelberg.

[Sei02]   Seidler, R. D., Alberts, J. L., & Stelmach, G. E. (2002). Changes in multi-joint performance with age. *Motor Control, 6*(1), 19–31.

[Shi05]   Shinar, D., Tractinsky, N., & Compton, R. (2005). Effects of practice, age, and task demands, on interference from a phone task while driving. *Accident; Analysis and Prevention, 37*(2), 315–326.

[Spi08]   Spink, A., Cole, C., & Waller, M. (2008). Multitasking behavior. *Annual Review of Information Science and Technology, 42*(1), 93–118.

[Str14]   Strobach, T., Salminen, T., Karbach, J., & Schubert, T. (2014). Practice-related optimization and transfer of executive functions: A general review and a specific realization of their mechanisms in dual tasks. *Psychological Research, 78*(6), 836–851.

[Tro07]   Trouvain, B., & Schlick, C. M. (2007). A comparative study of multimodal displays for multirobot supervisory control. In D. Harris (Ed.), *Lecture Notes in Computer Science. Engineering psychology and cognitive ergonomics* (vol. 4562, pp. 184–193). Berlin, Heidelberg: Springer Berlin Heidelberg.

[Vie15]   Vieluf, S., Godde, B., Reuter, E.-M., Temprado, J.-J., & Voelcker-Rehage, C. (2015). Practice effects in bimanual force control: Does age matter? *Journal of Motor Behavior, 47*(1), 57–72.

[Wah17]   Wahn, B., & König, P. (2017). Is attentional resource allocation across sensory modalities task-dependent? *Advances in Cognitive Psychology, 13*(1), 83–96.

[Wic80]   Wickens, C. D. (1980). The structure of attentional resources. In R. S. Nickerson (Ed.), *Attention and performance series. attention and performance VIII* (pp. 239–257). Hoboken: Taylor and Francis.

[Wic02]   Wickens, C. D. (2002). Multiple resources and performance prediction. *Jorunal of Theoretical Issues in Ergomoics Science, 3*(2), 159–177.

[Wis00]   Wishart, L. R., Lee, T. D., Murdoch, J. E., & Hodges, N. J. (2000). Effects of aging on automatic and effortful processes in bimanual coordination. *The Journals of Gerontology. Series B, Psychological Sciences and Social Sciences, 55*(2), 85–94.

[Woo02]   Woollacott, M., & Shumway-Cook, A. (2002). Attention and the control of posture and gait: A review of an emerging area of research. *Gait and Posture, 16*(1), 1–14.

[Wul10]   Wulf, G., Shea, C., & Lewthwaite, R. (2010). Motor skill learning and performance: A review of influential factors. *Medical Education, 44*(1), 75–84.

[Yan98]   Yan, J. H., Thomas, J. R., & Stelmach, G. E. (1998). Aging and rapid aiming arm movement control. *Experimental Aging Research, 24*(2), 155–168.

[Yog12]   Yogev-Seligmann, G., Hausdorff, J. M., & Giladi, N. (2012). Do we always prioritize balance when walking? Towards an integrated model of task prioritization. *Movement Disorders: Official Journal of the Movement Disorder Society, 27*(6), 765–770.

The Challenge of Being Self-Aware When Building Robots for Everyday Worlds

Andreas Bischof

Abstract Building robots to serve the needs of everyday life is described as a twofold challenge. Firstly, robotics, engineering, and computer science need new theories and concepts. Secondly, new methods and forms of collaboration are required. The article argues that the "wicked" nature of everyday worlds implies another, seemingly mundane challenge for roboticists: to become more self-aware about their own actions (Sect. 1). This includes a critical reflection on the goals of research and development, when dealing with humans (Sect. 2). Furthermore, there are two kinds of rather implicit methods roboticists use to make their machine work in everyday worlds, that should be considered more explicitly: On the one hand, the researcher's own everyday knowledge becomes an ambivalent resource for making decisions (Sect. 3.1). On the other hand, roboticists themselves often engage in creating expectations and desirable scenarios by staging robot behavior (Sect. 3.2). The article concludes not to wipe out these seemingly mundane practices, but rather to use their marginalization as a starting point for a reflective methodology of technical support systems (Sect. 4).

1 Everyday Worlds as "Wicked Problem" for Robotics

Dealing with 'real world problems' poses a number of challenges. The first and foremost is a crucial change within the object area of robotics: Humans and their interactions become part of the problem to make a robot work. Until the 1990s humans have been either a visionary reference point or a limiting condition for robotics. They were mostly considered as a safety risk. But the problems that arise when robots are deployed in everyday worlds cannot solely be understood as technical challenges, e.g., in terms of obstacle avoidance. Leaving the factory buildings and

A. Bischof (✉)
Junior Research Group "Miteinander", Media Informatics, University of Technology Chemnitz, Straße der Nationen 62, 09111 Chemnitz, Germany
e-mail: andreas.bischof@informatik.tu-chemnitz.de

© Springer Nature Switzerland AG 2018
A. Karafillidis and R. Weidner (eds.), *Developing Support Technologies*, Biosystems & Biorobotics 23, https://doi.org/10.1007/978-3-030-01836-8_12

127

laboratories does not only add a new set of tasks for the machines, it shocks the scientific—sociologists of science say: *epistemic*—foundations of the field.

Making robots work in everyday worlds does not only challenge the theories and methods of robotics, it also demands a new understanding of the role of the roboticist. By aiming at the actual use of robots in everyday life, robotics suddenly becomes a discipline such as architecture or urban planning, in which scientific, engineering, political, social, and aesthetic expertise meet. Robotics now shares the same resisting—some say malicious [Rit73]—kind of problem like architecture: Human activity in socio-technical systems is hard to operationalize and predict. Social situations and human(-robot) interactions are technically and scientifically incomplete. Factors and actions that become effective might not always be foreseeable. Demands of human-robot interaction are furthermore perspective-dependent, that is, they are subject to the interpretations of people—which often differ from the expectations of the designers and engineers. Furthermore, every action of a robot or a roboticist leads to reactions by the addressed users, so the roboticists expectations about the expectations of the future users may directly influence the resulting human-robot interaction.

Roboticists are forced to operationalize this social complexity into machine language. There is a stark contrast between the phenomenon of human-robot interaction in everyday worlds, which is difficult to standardize on the one hand, and robotics as an established set of problem-solving strategies, that rely on standardized procedures on the other hand. Other domains of Computer Science like software development or ubiquitous computing have related to the concept of "wicked problems" to describe this issue [DeG90, Cou08].

"Wicked problems" demand types of knowledge, skills, and perspectives that have not been part of robotics' self-concept, such as empathy, or bearing ambiguity and contingency of interpretation. Previous studies on robotics for everyday life have identified a lack of conceptual and practical approaches to deal with the "wickedness" of constructing robots for everyday worlds. The conceptual gap of robotics [Lin16] becomes obvious when compared to sociological and social-psychological perspectives. In summary, the criticism is that the (mostly implicit) social theories within robotics miss essential aspects of social interaction. Lindemann names the significant role of expectations for a successful human-robot interaction, and the indexicality of communication—both factors that are difficult to formalize [Lin16]. On a broader level, there is the strong tendency to conceptualize human-robot interaction as dyadic exchange between *two entities* on the micro level [Mei14, Hoe13]. This mostly cognitive approach to human-machine interaction, equivalent to "the second paradigm of HCI" [Har07], reduces the specific complexity of the everyday world: By focusing on cognitive factors, the complexity of the numerous, interacting, and context dependent social factors of everyday worlds become marginalized. To deal with this complexity is the core challenge for robotics in everyday worlds [Mei14].

In the following sections, I want to contribute to this challenge by discussing a methodological aspect of robotics research and development that is widely underexposed: the crucial role of robotics' and roboticists' own (inter-)actions towards and in everyday worlds. Their actions and interpretations do not only then become effective when a robot is deployed, but already in the very beginning of research and development. Therefore, I will argue for a *reflective stance within robotics* to the goals of research (Sect. 2), the everyday knowledge of the researchers (Sect. 3.1), and the way they foster expectations about robotic behavior (Sect. 3.2).

2 How Does Robotics Define Social Goals?

It is rarely publicly discussed, how goals and desired effects of robots in everyday worlds are defined. Research from Science and Technology Studies shows that the definition of goals in robotics research and development is mainly influenced by two factors: culturally shared "imaginaries" of robots on the one hand, and the conditions and constraints of research funding on the other hand. Šabanović's ethnographic study has shown that scientific theories and engineering problems are only a minor resource for the definition of objectives within social robotics. Moreover, everyday experiences, stories and symbols, especially from science fiction, played a decisive role and thus provided a common ground for researchers, the public, and funders [Sab07]. Studies comparing the culturally shared images of robots between Japan and the U.S. or Europe report similar findings [Lei06, Wag14]. The underlying concept of "imaginaries", collectively shared ideas that influence the researchers as well as the funding institutions, has a longer tradition in science and technology studies [Boe14]. It is known that imaginaries coordinate the communication of different groups of actors, e.g., the public and funding institutions [Roe08] and even the practices of researchers [Gie07]. It is characteristic that imaginaries interlink the technical feasibility and societal desirability of a technology, for example by promoting a politically desirable idea as being technically feasible to justify investments and funding [Jas09].

When looking at robotics research and development funding programs like the EU's SPARC [Noe14] or the National Robotics Initiative in the U.S. it becomes evident, that robotics funding is fueled by such claims of political desirability [e.g., Kro14]. Major research funding programs for technology development, for example in the field of Ambient Assisted Living, directly refer to this discursive figure. In these funding programs and the accompanying speeches, press releases, and workshops robotics is described as an instrument to control and compensate societal developments, like, e.g., the demographic change. This recurrent intertwining of solution promise and technological development has direct consequences for how robotics research and development sets its goals.

The classical epistemological path of producing knowledge in science is thus reversed. Robotics orientation of research and development can be called "post hoc" [Kno84]. The aim of this kind of work is not so much to discover *new solutions*, as rather to successfully *implement the previously defined solution* "robot application", as Meister has pointed out for service robotics [Mei11]. Whether these objectives defined upfront are actually relevant for everyday worlds is most often not a superior criterion to decide upon the goals of robotics research and development.

The most problematic aspect of this framing is, that its effect on the actual research and development practices becomes implicit. The instrumental reference to everyday worlds as "context of application" makes it invisible, how users and usage scenarios are configured by upfront defined objectives [Woo90]. This blind spot is methodologically and ethically problematic. If robotics does not explicitly reflect and discuss the implications of its goals, it runs the risk of being a self-fulfilling prophecy of the imperative of usefulness, rather than building robots that actually fit in everyday worlds.

The following questions provide a rough guide to reflect on the origin and function of goals in robotics research and development and can easily be applied to any kind of project:

- *Who* defined the goal of R&D/task for the robot?
- *When* was the goal defined—before or after a contact with the users addressed?
- Is there any valid *empirical evidence*, that this goal refers to actual needs of users in everyday worlds?

3 Silent Contributions by the Researchers and Engineers

Human-robot interaction and other branches of robotics adapted a plethora of methods and methodological approaches to research everyday worlds and users' interactions with machines. Methods like ethnographic observations of field tests, in-depth interviews with users or acceptance tests with focus groups are particularly suitable to deal with the "wicked" nature of building robots for everyday worlds. More quantitative, lab-oriented methods like usability or task tests in closed environments, or questionnaires on perceived quality of human-robot interaction most often do not lack explicit reflection either. Most of these methodological applications from fields like psychology, anthropology or even design research evolved to a comparably high standard of methodical reflection.

However, there is an unseen area of robotics research and development, that is only unsatisfactorily controlled or reflected methodologically: roboticists actions towards their research objects and subjects. Thereby I mean two sets of activities that are not considered scientific or academic at all, yet have a great impact on the demands and expectations towards robots in everyday worlds. The first are everyday activities and everyday knowledge of the researchers that implicitly become a resource for understanding and defining demands (Sect. 3.1). The second are activities of staging robot behavior and thereby actively creating expectations of robots' capabilities (Sect. 3.2).

3.1 Everyday Knowledge as Ambivalent Resource

It should come as no surprise that everyday knowledge is a resource to make robots work for everyday worlds. Relying on one's own observations, implicit knowledge, and incorporated abilities in order to achieve a fit of machines and everyday scenarios is an obvious and promising strategy. By doing so roboticists become instruments of robotics requirement engineering. They draw on experiences and expertise gained as everyday people. This has been confirmed in many cases during my visits in robotics laboratories in the U.S. and Europe [Bis17]. Particularly when talking about the motivation and aptitude to work in the field of social robotics it becomes obvious, that everyday occurrences in the lives of researchers have an epistemic value for their work. However, these everyday methods and this knowledge are most often neither documented nor questioned, which is problematic.

An example for such everyday methods of researchers is "lay ethnography". Thereby I mean observations of everyday life done by roboticists. For example, a researcher followed people through a university building in order to find out which floors the students and staff typically use and what goals they have. Although the roboticist does not consider such an activity to be a proper scientific experiment, it still helps him or her to limit the space of solvable problems in order to determine the further course of the robotic project. (The specific case retold here aimed at building a service robot that fulfils simple tasks in the university building.) The strategy applied by the researcher is typical for laymen's ethnographies in robotics: "Let's just see how people do it."

These lay ethnographies differ from methodically controlled ethnographies in their purpose: Lay ethnographies are means of generating meaning within the researchers' work [Wee06]. They are rarely discussed outside of a laboratory nor presented at conferences or contested by competing interpretations or further empirical material: lay ethnographies only serve themselves so to speak. They are not aimed at being transformed into a discourse that explains and contrasts the researcher's own understanding, as professional ethnographies are required to do. Without this reflective element, everyday observations of researchers miss out the core analytic quality of ethnography [Dou06]. Lay ethnographies are a rather descriptive method, that does not question the point of view of the observer. But this would be highly critical for an ethnography in order to be valid. There is a long tradition in Science and Technology Studies analyzing and criticizing such "I-Methodology" of engineers and designers [Akr92], that consist in extrapolating knowledge from their point of view.

3.2 Creating Expectations About Robots

Most robotics research teams develop a kind of "demonstration routine". There are plenty occasions, were roboticists are asked to present their work, such as open house days for publics or funders, during events to attract new students, competitions or Science Fairs. Due to these regular presentations of their machines, most roboticists gain a feeling for the expectations of the audience, the effects of certain behavioral patterns, and also for the necessity of props and scripts.

The "YouTubization" of research [Bot15] is an expression of this important role of staging robots' behavior in robotics. The creation and circulation of "demo videos" of robots are almost obligatory for robotics research. For a "demo video", the behavior of the robot is usually presented in a scripted scenario in which a human demonstrator or narrator comments on the machine. These presentations serve to illustrate the feasibility of a technical solution or the fault-free operation of a prototype [Ros05]. Such videos are omnipresent in robotics: they can be found on the researchers' websites, are used in lectures, in conference talks, become part of the publications, and circulate on YouTube, Facebook, and in tech blogs as well as in the researchers' mail inboxes.

This popularity has inspired a growing body of research literature on the role of demo videos for research and development [Ros05, Suc07, Suc11, Suc14, Win08, Bot15]. Winthereik et al. [Win08] examined the staging of future uses in such videos. They found that the videos first and foremost serve the "life-world of the demo", the places and occasions of the demonstration of the videos: Above all they are a tool for connecting to discourses of potential industrial cooperation partners, marketers, politicians, colleagues, and user groups. Both highlighted the expressional element in the researchers' activity of staging robotic behavior along the aesthetic conventions of video clips [Bot15]. Demo videos are thus also an expression of the identity constructions of the researchers, but this does not mean that they depict themselves in the videos. Instead, the focus is almost entirely on the performance of the machines. This can be interpreted as a successful representation of the original goal of new robotics: to build machines that are suitable for everyday use. The videos and their productions then aim to provide proof of the robots' efficiency.

These practices of staging include clearly an interaction between the researchers and the addressed users. "Demo videos" are shaped by expectations and form expectations for human-robot interaction. Robotics does not just rely on cultural imaginaries of robots' capabilities, it shapes them too. The staging of technically not yet feasible robot behavior is part of the negotiation on the potential of a technology and for what purposes it might be used [Win08, Lat05]. Many researchers in the field are well aware that they contribute to a changing experience of sociality by developing their machines and creating presentations of them [Dau98, Tur17]. However, the implications and conditions of this change—and the roboticists' share thereof—are not explicitly reflected and discussed in robotics research and development.

4 Conclusion: Towards a Reflective Methodology of Robotics for Everyday Worlds

In this text, I pointed towards some critical practices of roboticists. However, I did not do this to discredit or de-legitimize the goal to build autonomous machines or support systems. Instead, the empirical findings of Science and Technology Studies and Sociology of Science are meant to improve the ability of engineers and scientists to include themselves in the equations they make about everyday worlds. Engineering robots for everyday worlds has been described as a "wicked problem". This means that the objects in this area are so complex and changing that there are no standardized solutions to treat them. Instead, the processing of "wicked problems" essentially *depends on the formulation and definition of the problem by the developers themselves* [Rit73].

Beside psychological theories and effects, robotics for everyday worlds should focus on the dynamics of social worlds and the diversity of perspectives of different stakeholder groups—including roboticists themselves. New theories and methods require a different understanding of the relationship between researchers and their 'objects' of investigation, which are alive, interpreting, responding, and involved.

The proposal of the article is to take a reflective stance on practices and assumptions within research and development—even when they seem mundane and/or marginal to the goal of making the robot work technically. As we have seen, within everyday worlds the fit between man and machine goes far beyond technical problems. It is also a matter of culturally shared meaning, contesting definitions of usefulness, and power with respect to defining goals. Thus, making robots work in everyday worlds requires heuristics that go far beyond 'neutral' measurements and technical modeling. The personal contribution of roboticists cannot be ignored nor eliminated from the task to build robots for people.

This extends to all creators of technical support systems, although their specific technical and social challenges may differ. Overarching fields like Human-Computer interaction or concrete domains like wearable technologies or smart homes share the problem to mechanize and digitize practices and situations that are *wicked* in the above described sense. Thereby engineers and designers of support systems have been assigned a leading role in the overall process and not only with regard to the construction of the artifact proper: Their products embody social relations, for example power relations, and they are responsible to reflect upon this. This transcends simple categories like 'intended' or 'unintended' effects of technology use. It requires to critically consult the technical development as a process and to understand, whether it is biased in a particular direction, or which social interests it favors. The process of technological development is critical in determining the politics of an artifact [Win80]; hence the importance of incorporating all stakeholders—instead of viewing the actions and beliefs of scientists and engineers as somehow external to the social and normative dimensions of support systems.

References

[Akr92]   Akrich, M. (1992). The de-scription of technical objects. In W. E. Bijker & J. Law (Eds.), *Shaping technology/building society. Studies in sociotechnical change* (pp. 205–224). Cambridge: MIT Press.

[Bis17]   Bischof, A. (2017). *Soziale Maschinen bauen: Epistemische Praktiken der Sozialrobotik*. Bielefeld: Transcript.

[Boe14]   Böhle, K, & Bopp, K. (2014). What a vision: The artificial companion. A piece of vision assessment including an expert survey. *Science, Technology & Innovation Studies (STI Studies), 10*(1), 155–186.

[Bot15]   Both, G. (2015). Youtubization of research. Enacting the high-tech cowboy in video demonstrations. In S. Davies, M. Horst, & E. Stengler (Eds.), *Studying science communication* (pp. 24–27). Bristol University of the West of England.

[Cou08]   Courtney, J. F. (2008). Decision making and knowledge management in inquiring organisations: Toward a new decision-making paradigm for DSS. *Decision Support Systems—Knowledge Management Support of Decision Making, 31*(1), 17–38.

[Dau98]   Dautenhahn, K. (1998). The art of designing socially intelligent agents: Science, fiction, and the human in the loop. *Applied Artificial Intelligence, 12*(7–8), 573–617.

[DeG90]   DeGrace, P., & Stahl, L. H. (1990). *Wicked problems, righteous solutions: A catalog of modern engineering paradigms*. Yourdon Press.

[Dou06]   Dourish, P. (2006). Implications for design. In *Proceedings of the SIGCHI Conference on Human Factors in Computing Systems* (pp. 541–550). New York: ACM.

[Gie07]   Giesel, K. (2007). *Leitbilder in den Sozialwissenschaften*. Wiesbaden: Springer.

[Har07]   Harrison, S., Tatar, D., & Sengers, P. (2007). The three paradigms of HCI. *Alt. Chi. Session at the 2007 SIGCHI Conference on Human Factors in Computing Systems* (pp. 1–18). New York: ACM.

[Hoe13]   Höflich, J. R. (2013). Relationships to social robots: Towards a triadic analysis of media-oriented behavior. *Intervalla, 1*, 35–48.

[Jas09]   Jasanoff, S., & Kim, S. (2009). Containing the atom: Sociotechnical imaginaries and nuclear power in the United States and South Korea. *Minerva, 47*(2), 119–146.

[Kno84]   Knorr Cetina, K. (1984). The fabrication of facts: Toward a microsociology of scientific knowledge. In M. Stehr (Eds.), *Society and knowledge* (pp. 223–244). Oxford: Transaction Books.

[Kro14]   Kroes, N. (2014). Lighting a SPARC under our competitive economy. Speech available at http://europa.eu/rapid/press-release_SPEECH-14-421_en.htm.

[Lat05]   Latour, B., & Weibel, P. (Eds.). (2005). *Making things public: Atmospheres of democracy*. Cambridge: MIT Press.

[Lei06]   Leis, M. J. (2006). *Robots–our future partners. A sociologist's view from a German and Japanese perspective*. Marburg: Tectum.

[Lin16]   Lindemann, G. (2016). Social interaction with robots: Three questions. *AI & Society, 31*(4), 573–575.

[Mei11]   Meister, M. (2011). *Soziale Koordination durch Boundary Objects am Beispiel des heterogenen Feldes der Servicerobotik*. Doctoral Thesis, Fakultät Planen, Bauen, Umwelt, Technische Universität Berlin.

[Mei14]   Meister, M. (2014). When is a robot really social? An outline of the robot sociologicus. *Science, Technology & Innovation Studies, 10*(1), 107–134.

[Noe14]   Nördinger, S. (2014). Ziviles Forschungsprogramm SPARC. EU-Komission startet weltweit größtes ziviles Robotik-Forschungsprogramm. http://sparc-robotics.eu/wp-content/uploads/2014/06/Produktion-Germany-EU-Komission-startet-weltweit-gr%C3%B6%C3%9Ftes-ziviles.pdf.

[Rit73]   Rittel, H., & Webber, M. (1973). Dilemmas in a general theory of planning. *Policy Sciences, 4*(2), 155–169.

[Roe08]   Roelofsen, A., Broerse, J. E., de Cock Buning, T., & Bunders, J. (2008). Exploring the future of ecological genomics: Integrating CTA with vision assessment. *Technological Forecasting and Social Change, 75*(3), 334–355.

[Ros05]   Rosenthal, C. (2005). Making science and technology results public: A sociology of demos. In B. Latour & P. Weibel (Eds.), *Making things public*. Cambridge: MIT Press.

[Sab07]   Šabanović, S. (2007). *Imagine all the robots: Developing a critical practice of cultural and disciplinary traversals in social robotics*. Doctoral Thesis, Rensselaer Polytechnic Institute.

[Suc07]   Suchman, L. (2007). *Human-machine reconfigurations: Plans and situated actions*. Cambridge: University Press.

[Suc11]   Suchman, L. (2011). Subject objects. *Feminist Theory, 12*(2), 119–145.

[Suc14]   Suchman, L. (2014). Humanizing humanity. Blogpost: https://robotfutures.wordpress.com/2014/07/19/humanizing-humanity/.

[Tur17]   Turkle, S. (2017). *Alone together: Why we expect more from technology and less from each other*. Hachette.

[Wag14]   Wagner, C. (2014). Techno-imaginations and robot role models: Discussing the influence of popular culture on the development of next generation robots in Japan. In *Proceedings of the 2nd IEEE International Conference on Universal Village*, June 16–17, 2014, Boston.

[Wee06]   Weeks, J. (2006). Lay ethnography and unpopular culture. Working Paper 2006/47/OB, 2006.

[Win08]   Winthereik, B., Ross, N., & Strand, D. L. (2008). Making technology public: Challenging the notion of script through an e-health demonstration video. *Information Technology & People, 21*(2), 116–132.

[Win80]   Winner, L. (1980). Do artifacts have politics? *Daedalus, 109*(1), 121–136.

[Woo90]   Woolgar, S. (1990). Configuring the user: The case of usability trials. *The Sociological Review, 38*(1), 58–99.

Engineering Collaborative Social Science Toolkits. STS Methods and Concepts as Devices for Interdisciplinary Diplomacy

Peter Müller and Jan-Hendrik Passoth

Abstract The smartification of industries is marked by the development of cyber-physical systems, interfaces, intelligent software featuring knowledge models, empirical real-time data, and feedback-loops. This brings up new requirements and challenges for HMI design and industrial labor. Social sciences can contribute to such engineering projects with their perspectives, concepts and knowledge. Hence, we claim that, in addition to following their own intellectual curiosities, the social sciences can and should contribute to such projects in terms of an 'applied' science, helping to foster interdisciplinary collaboration and providing toolkits and devices for what we call 'interdisciplinary diplomacy'. We illustrate the benefits of such an approach, support them with selected examples of our involvement in such an engineering project and propose using methods as diplomatic devices and concepts as social theory plug-ins. The article ends with an outlook and reflection on the remaining issue of whether and in how far such 'applied' and critical social science can or should be integrated.

1 Social Science in Engineering Projects

The transition from traditional to smart industries is indicated by the implementation of (digital) automation tools and the integration of cyber-physical systems, e.g., sensory interfaces, informed models and (self-learning) algorithms. This refurbishment of factory (infra)structure is affecting HMI design decisions and the ways that such smartified plants are operated [Pos15]. Industrial smartification thus involves dense entanglements of business, organization, technology and labor-routine issues, all of which are of genuine interest to disciplines like sociology and "Science &

P. Müller · J.-H. Passoth (✉)
Munich Center for Technology in Society/Digital Media Lab, Technical University of Munich, Arcisstraße 21, 80333 München, Germany
e-mail: jan.passoth@tum.de

P. Müller
e-mail: pet.mueller@tum.de

© Springer Nature Switzerland AG 2018
A. Karafillidis and R. Weidner (eds.), *Developing Support Technologies*, Biosystems & Biorobotics 23, https://doi.org/10.1007/978-3-030-01836-8_13

Technology Studies" (STS). In the context of a deep involvement in an engineering project on smart factories and industrial automation, we as STS researchers have composed a qualitative methods toolbox to help understand the effects of a transition from traditional to smart production. In Sect. 3 of this paper, we will give a very short overview of this toolbox and how we used it in the project. We did this not only for the sake of a sociological understanding (of socio-technical arrangements) but also for engineering purposes of human machine interface (HMI) design as well as to acquire technological, formal knowledge models. Similar methodologies have already been designed and elaborated in specific engineering disciplines, especially by those who work in the field of HMI [Jan16]. In HMI, but also other engineering fields, methods from cognitive science and ergonomics and also from the social sciences (in a way, since they are socio-scientific methods being used for different purposes) are commonly employed in order to acquire (expert) knowledge, develop mental models and assess suitable and supportive interface designs [Liu10]. By providing a toolbox, we further contribute by adopting sociological methods not limited to cases in which social science project members demand such a methodology. We will also illustrate how such a methodology might contribute to the methodological repertoire of this specific field of engineering, turning our methods into devices for interdisciplinary diplomacy.

To illustrate the usefulness of such an approach, we used this toolbox and a set of STS concepts to develop recommendations for the training of industrial plant operators using several generic training scenarios designed specifically for smart factory HMI cases. We will highlight some of these recommendations in Sect. 4 of this paper. Our goal was to come up with a broader framework to constantly revise and update the training scenarios that have already been applied and also to train supervising staff in how to use the data generated thus far to sort out common patterns of successful plant operation that go beyond intrinsic plant models and practical operating guidelines like component maintenance. The purpose of such an approach is to help guide the reconfigurations of socio-technical arrangements in complex work environments that are at the core of transitions from traditional to smart production. Such a transition, we conclude, must be careful—following Annemarie Mol's notion of care [Mol08] in the sense that it needs care and must be taken care of—and we must be incremental to ensure it is done in a responsive and responsible way that helps to increase both the autonomy and the self-determination of operators and supervising staff.

We will conclude this piece with a short reflection (Sect. 5) on the affirmative or critical nature of such an approach to the integration of social science research in engineering projects. We will argue that while it is true that such deep integration in projects makes it tricky to criticize their overall ends, it is only through such immersion that it is possible to produce concrete alternatives and challenge taken-for-granted assumptions in engineering.

2   Smart Factories and Smart Collaborations

As partners of the EU-funded project 'IMPROVE', we investigated the 'socio-technical aspects' of the project. IMPROVE is focused on developing automation technology for smart factories, improving plant surveillance for operators, and providing self-learning software that contains plant models and processes real-time data which are used for early anomaly detection and malfunction anticipation. But how do we as social scientists and STS researchers function as a part of this project? We worked very close with the HMI engineers from IMPROVE who were not only commissioned with designing an interface for surveillance and detection features, but also with developing a decision support system (DSS) for plant operators in this and other smart factory contexts. Furthermore, their work package involved the task of eliciting expert informed mental models of the industrial project partners' plants. To do this, they used card-sorting techniques and interview sessions with the technical and operating personnel from IMPROVE's industrial partners. From this they were able to build ordinal cause-effect graphs to be used for operating purposes in terms of monitoring and decision support. For our deliverable as social science partners, it was necessary to collaborate with the HMI project team in order to understand their design assignments and means and thus to contribute by providing socio-scientific consulting and complementary content like data or concepts. Our engineering partners shared their data (and resulting models and prototypes) and took us with them to their inquiry meetings. Based on these data and experiences we attempted to understand the capacities and underlying strategies of their methodology—an eclectic assemblage of psychological, ergonomic, and social science methods. Furthermore, we could gather our own field data—on the investigated operators and our researcher colleagues and their very own social practices of engineering [Buc94]—and experimented with ways of analysis that could complement our partners' data analysis. While they used the data to create the plants' models, we have tried to figure out some characteristics of operators' labor routines that would be relevant to HMI design. As far as they could be reconstructed based on our dataset, we drafted concepts of how operators and supervisors handle problems, how they conceptualize their own practice, and how they configure their organizational roles.

These very socio-technical arrangements that we have mapped informed our additional tasks involving the development of a concrete, complementary toolkit of qualitative research methods that would fit such engineering projects. Hence, we not only acted as methodological consultants but, in particular, as methodological researchers who were tinkering with a particular toolbox that could inform HMI design and DSS features. Also, we experimented with generic training scenarios for operators in smart factories in ways that featured IMPROVE's or similar technology. We grounded both assignments in the same data and concepts in order to design an analytical setup for our toolbox and in order to develop training scenarios that took into account the operators' particular practices and the specific knowledges and skills that are required. Therefore, we have worked on a conceptualization for practice blueprints concerning the different types of problems operators encounter in their work. Our idea was that the HMI design and practices covered by the training schemes provided should be more integrated and could even feature reciprocal synergies.

We have also observed another differentiation of methodological research approaches that is seen less frequently in sociological studies. In addition to the qualitative methods applied by IMPROVE's HMI partners, other partners used data mining to receive quantified, explicit data and models that would feature precise predictions. The applied quantitative and qualitative methods of our engineering partners were decidedly focused on explicit correlations and configurations of data and models that were both quantitative and qualitative. Taking our own contribution into account, this adds up to a threefold (at least) methodological setup, where we provided research methods that mostly covered implicit, tacit knowledge and latent, subtle orders of practice. While our partner oriented their work towards use-cases, we were rather looking for 'problem-cases', e.g., unresolved tensions of interest, places of interference between engineering and worker (operator) mindsets. We prepared a methodological toolbox because we wanted not only to contribute additional data as social science experts but also to add a perspective that could cross and integrate our partners' quantitative and qualitative approaches and findings. To do so, it was necessary to re-think our roles as methodologists: (how) could we (re)invent ourselves as socio-engineers of diplomatic devices within interdisciplinary engineering projects?

3   Methods as Diplomatic Devices

Qualitative research methods are already known within the fields of engineering, especially digitalization. However, from a sociological point of view, these methods are not sufficiently elaborated—or at least not properly implemented and justified—by their users. In many publications that deal with such methodological needs or issues, corresponding methods are applied but not explained. In Software Engineering, this has been recognized and initial steps have been explicitly taken to turn towards more sound empirical research [Dit08, Tor11, Han07], but such an approach is missing from industrial engineering. Nonetheless, this must not be mistaken for some kind of sociological snobbism. This is not about pushing sociological questions into other disciplines, but clearing up what such qualitative methods are capable of, and how their application can even be used to help produce an understanding of disciplinary boundaries and ways of crossing them. As a toolbox for collaborative projects, they can be thought of as *diplomatic devices*. We have turned several social science methods into such devices. In particular, we have focused on the qualitative analysis of technical documents and on ethnographic fieldwork. A textual version of this toolbox has been created for our project deliverable and will soon be published as part of collaboratively edited collection [Pas18].

In engineering contexts, these methods—although in most cases it is only interviews—are used as knowledge-acquiring methods that are focused on objectives. These methods are, or at least seem to be, rather formalized, explicit and objective (insofar as they describe objects' qualities). As a result, such methods mostly consist of cognition science methods, ergonomics, and (complementary) interviews [Han07, Jan16]. These interviews, in particular, seem to back up the other methods of model generating and literary review. Interviews, however, are not a standard procedure only within the humanities and social sciences. The sociological style of doing interviews, however, is more elaborate in certain respects. This concerns several methodological principles and guidelines, e.g., CA transcription (conversation analysis transcript or CAT) [Sch73] or the several interview guideline revisions (whether it is open, closed or structured) due to their pretesting results. From an engineering perspective, these methodological norms might seem very exaggerated. Furthermore, since engineering assignments were met with less methodological effort when it comes to using interviews for requirements engineering and testing purposes, one might be tempted to agree with this assessment. However, we will give several arguments to the effect that these efforts do, in fact, pay off scientifically and in terms of technology development and design. Integrating social science researchers and social science methodology enables a project to ground its work in a common understanding whose quality satisfies all its epistemological and technical requirements and which is common because it has been established collaboratively. For example, it is possible to ground the engineering of formal models, HMI and socio-scientific reflections on the same empirical data, thus increasing the capacity for (and likelihood of) synergetic exchange between different researchers and of the overall coherence within such a project.

In the course of the project, we were able to accompany our engineering colleagues on requirements engineering visits, provide methodological consultation and host data analysis sessions. We attempted to turn the methodological canon and controversies of social science interview research into a toolbox equipped with a heterogeneous set of devices, we assumed that the organizing principle of such a collection of devices does not need to be rigor and coherence, but rather its usefulness. It was designed to be useful for the common project of treating the various actors we encountered (operators, managers, industrial researchers and, yes, ourselves) as part of a 'public' as John Dewey understood it [Dew06]: as something that "cannot be mastered by anyone but that can be represented, over and over again, by the social sciences and the humanities" [Lat03]. This might also be identified as a Meadian institution, for it addresses "situations which we admit are not realized but which demand realization" [Mea23], thus are to be handled constantly in an infinite struggle of methodological feedback. This toolbox—a collection of revisable how-tos, visits and workshops—was the basis for our own substantial contribution to the overall project (see Sect. 5 for a short reflection on this). It was also the basis for unpacking a controversy dealing with claims of validity, epistemic authority and pragmatic usefulness instead of just glossing over the differences between disciplinary cultures by proposing a methods-based consensus.

4   Social Theory Plug-ins

As partners in the project consortium, our task as STS researchers was not only to encourage interdisciplinary research and help our engineering partners talk to actors they encounter in the empirical context of industrial research and practice, but also to provide recommendations on how to deal with the reconfiguration of sociotechnical arrangements that are at the practical core of the transition from traditional to smart production. Our recommendations for dealing with the reconfiguration of sociotechnical arrangements are prototypes because the work of our engineering partners, for example, on the use of machine learning for semi-automatic alarm analysis or on the prototype for a self-adapting HMI, are also proof-of-concept demonstrators and prototypes rather than concrete implementations of a new version of a marketable production facility. We explicitly used three *social theory plug-ins*: a relational and procedural concept of agency based on pragmatism [Dew96, Mea03], social science approaches to implicit knowledge based on practice theory [Col01] and a conceptual framework for human-machine cooperation based on Actor-Network Theory [Pas15]. They provided the ground for our design and training recommendations, we used them to challenge and rework the mostly implicit, but sometimes also very explicit, assumptions about work and automation, tacit knowledge, and HMI principles used for modeling and design by our engineering partners. As in the case of the toolbox described above, these plug-ins served a double purpose for us: they are at the same time provisional results of STS research in an engineering project as well as ways of intervening in the daily work of engineering practice. In this way, they open up already closed (and sometimes too quickly closed) debates about goals and work packages and propose alternatives and collaboratively develop ways of dealing with the changes in sociotechnical arrangements introduced by the project as a whole [Jen01]. They are "lateral concepts" that enable "ontological experiments" [Gad16, Jen15].

One example of this approach is the following: By using these *social theory plug-ins*, we argued, in design meetings and in comments on requirements and models, that HMIs should not be regarded as one-way tools for monitoring and controlling machines because operators will then be required to solve the (nearly impossible) problems of translating their implicit knowledge into explicit machine instructions and of mapping (standardized) HMI features onto (tacitly) known routines and patterns of trained behaviour. On the contrary, we suggested that operators should at least be able to organize themselves through the HMI and reflect upon the HMI's role in their practices in case of suboptimal operating processes. This is no mere rejection of operator responsibility, which, since it plays a significant role in daily work should still remain on the operator's side. Rather, it enables a different way of accounting for best practices, work-arounds, glitches and failures in any current and future HMI design. However, operators do not just use HMI. Training can focus on supporting operators to enhance themselves through the HMI, which will also help to resolve responsibility and decision dilemmas in case an integrated decision support system contradicts the operator's intuition. We therefore suggested that operator trainings need to focus on three issues: a symmetrical approach to HMI that enhances human-machine cooperation and adaptation in both directions, classifying incidences by whether they require either implicit or explicit knowledge and managing their (and other operators') knowledge and experience through the HMI by giving feedback on provided information and recommended interventions as well as by reporting and storing their own knowledge and experiences. Social theory of knowledge in practice and actor-networks can thus be used to organize information, classify incidences and provide an analytic framework for further, recursive adaptations [Ber98].

5 Involvement and Intervention

The interdisciplinary approach we have presented here goes beyond the mere contribution of extra-technological contexts like marketing, organization, policy, technology assessment or 'nice to have' contemplations. To reflect on the implicit premises of technological developments or on the societal meaning of their implementation is usually regarded as a genuine and specifically sociological duty. While, from an engineering point of view, such contemplation might rather appear as an ornament of interdisciplinary projects, it is exactly what social scientists regard as their primary obligation, thus those often harshly criticize colleagues who engage with instrumental, affirmative tasks in such interdisciplinary projects for doing so.

But are 'applied' and reflexive social science approaches, after all, mutually exclusive options [Hor02]? Collaborations between social scientists and engineers require the social scientist, indeed, to get her- or himself into the technical, instrumental setup of the engineering project—and that means, at least for the sake of cooperation, accepting the project's frame of reference and affirming the project's cultural, economic and organizational premises, established facts and assumptions. As a result, social science critiques are constrained by the explicit ends defined by the project, and radical critique and reflections on the *conditions of possibility* of such social situations are rendered impossible. Although social scientists do not need to completely narrow their perspectives, concerns, and issues regarding the project's framework, collaboration might yet compromise their capacities for social criticism. This is a concern that causes many social scientists to reject 'applied' science scenarios of social science. It is, of course, quite possible for social scientists to add subaltern interests and critiques to the tenor of what is taken into account concerning technological design. For example, by reflecting on diversity and thus inspiring more inclusive interfaces or devices (e.g., airbags positioned to consider physiognomic gender disparities) or helping to design software that incorporates organizational responsibility and accountability distributions or that helps to eliminate hierarchical tensions (e.g., bi-directional communication or feedback loops instead of mere monitoring and intervening). In this way, immediate problems are resolved, but the structural sources of these problems remain intact and might even (re)appear in a hardened form. After all, collaboration can also be a pejorative term, and it thus holds on to this residual connotation despite its recent popularity in terms of interdisciplinarity [Nie14].

To conclude, is the integration of applied and critical social science designed to fail? It is unquestionable that, however such integration is done, it can neither avoid trade-offs (on the instrumental or critical side) nor replace proper, exclusive social criticism. Nevertheless, interdisciplinary collaborations between social scientists and engineers have two advantages: on the one hand, they offer a deep insight into engineering cultures, in terms of engineers' working culture and the cultural significance of technological artifacts with respect to their demands and effects. On the other hand, such collaborations enable social scientists to provide interdisciplinary diplomacy, to mediate and offer consultation within their project and with regard to its social context. Without forgetting about the aforementioned critique conundrum, both features meet certain aspirations of social criticism: to get involved and be in touch with social situations, to contribute tangible and actual (critical) interventions. This avoids the separation of practical and intellectual work corresponding to social segregation and stratification [Hor02]. Eventually, if "technology is society made durable" [Lat90], the usual critical practice of watching and judging from afar is more than unacceptable because the question of which society is made durable and how is a never-ending, substantial concern. Interdisciplinary diplomacy is a way to share and spread this concern within interdisciplinary collaborations and to start working (together) on concrete alternatives.

References

[Ber98] Berg, M. (1998). The politics of technology: On bringing social theory into technological design. *Science, Technology and Human Values, 23*(4), 455–491.
[Buc94] Bucciarelli, L. L. (1994). *Designing engineers.* MIT Press.
[Col01] Collins, H. M. (2001). What is tacit knowledge? In: Schatzki/Cetina/Savigny (Eds.), *The practice turn in contemporary theory* (pp. 115–128).
[Dew96] Dewey, J. (1896) The reflex arc concept in psychology. *Psychological Review 3, American Psychological Association*, (pp. 357–370).
[Dew06] Dewey, J. (2006). *The public and its problems.* USA: Ohio University Press.
[Dit08] Dittrich, Y., Rönkkö, K., Eriksson, J., Hansson, C., & Lindeberg, O. (2008). Cooperative method development. *Empirical Software Engineering, 13*(3), 231–260.
[Gad16] Gad, C., & Jensen, C. B. (2016). Lateral concepts. *Engaging Science, Technology, and Society, 2,* 3–12.
[Han07] Hannay, J. E., Sjoberg, D. I. K., & Dyba, T. (2007). A systematic review of theory use in software engineering experiments. *IEEE Transactions on Software Engineering, 33*(2), 87–107.
[Hor02] Horkheimer, M. (2002). Traditional and critical theory. In: *Ibid: Critical theory. Continuum* (pp. 188–243).
[Jan16] Jander, H., Borgvall, J., & Ramberg, R. (2016). Towards a methodological framework for HMI readiness evaluation. *Human Factors and Ergonomics, 56,* 2349–2353.
[Jen01] Jensen, C. B. (2001). CSCW design reconceptualised through science studies. *AI & Society, 15*(3), 200–215.
[Jen15] Jensen, C. B., & Morita, A. (2015). Infrastructures as ontological experiments. *Engaging Science, Technology and Society, 1,* 81–87.

[Lat90]   Latour, B. (1990). Technology is society made durable. *The Sociological Review, 38*(1), 103–131.

[Lat03]   Latour, B. (2003). Is re-modernization occuring—And if so, how to prove it?: A commentary on Ulrich Beck. *Theory Culture and Society, 20*(2), 35–48.

[Liu10]   Liu, Y., Osvalder, A.-L., & Karlsson, M. A. (2010). Considering the importance of user profiles in interface design. In: Mátrai (Ed.), *User interfaces* (pp. 61–80). Intech.

[Mea23]   Mead, G. H. (1923). Scientific method and the moral sciences. *International Journal of Ethics, 33,* 229–247.

[Mea03]   Mead, G. H. (1903). The definition of the psychical. *Decennial Publications of the University of Chicago, 3,* 77–112.

[Mol08]   Mol, A. (2008). *The logic of care: Health and the problem of patient choice.* London: Routledge.

[Nie14]   Niewöhner, J. (2014). Perspektiven der Infrastrukturforschung: carefull, relational, kolobariv. In Lengersdorf and Wieser (Eds.), *Schlüsselwerke der Science and Technology Studies. Transcript* (pp. 341–353).

[Pas15]   Passoth, J.-H. (2015). Heterogenität und die Hybriden: Die Unbestimmtheiten der Actor-Network Theory. In Kron (Ed.), *Hybride Sozialität-Soziale Hybridität.* Velbrück Wissenschaft (pp. 89–108).

[Pas18]   Passoth, J.-H., & Müller, P. (forthcoming). Socio-technical arrangements of smart factory HMI. In Schüller and Niggemann (Eds.), *SpringerOpen IMPROVE Special Issue.* Berlin: Springer.

[Pos15]   Posada, J., Toro, C., & Barandiaran, I. (2015). Visual computing as a key enabling technology for Industrie 4.0 and industrial internet. *IEEE Computer Graphics and Applications, 35*(2), 26–40.

[Sch73]   Schegloff, E. A., & Sacks, H. (1973). Opening up closings. *Semiotica, 8,* 289–327.

[Tor11]   Tore, D., Prikladnicki, R., Rönkkö, K., Reaman, C., & Sillito, J. (2011). Qualitative research in software engineering. *Empirical Software Engineering, 16*(4), 425–429.

Part III
Forms and Contexts of Deployment

Development projects of support technologies are more than any other form of technology construction contingent on the early deployment of prototypes. This is, as it were, the feedback that the technical systems give to the researchers and developers—in addition to the feedback of the users and other stakeholders. Prototypes then might grow into products or are supplanted by other prototypes. All of them, however, get tested in specific environments and spawn insights about components and materials suitable for an integration into the human worlds of life and work.

Part III of this volume therefore presents various contexts and also possible forms of deployment. "Forms" of deployment address the concrete engineering of specific components that are pertinent for developing technologies and qualify for an integration into human activities (e.g., soft robotics, wearable sensors, pneumatic actuators). "Contexts" of deployment refer to the concrete domains of action and communication in which relevant support systems are embedded, tested, and optimized.

One major deployment context for support technologies is currently industrial production. This part of the book therefore starts with a contribution of *Robert Weidner, Bernward Otten, Andreas Argubi-Wollesen*, and *Zhejun Yao* who give an overview of typical challenges of industrial work and how these can be met with pertinent support systems. They discuss respective exoskeletal solutions that have proved themselves in practice or seem promising for further development.

Subsequently, *Yves-Simon Gloy* gives a concise description of the development of an augmented reality support system in the textile industry. He shows how existing features of a weaving machine are combined with a support system designed to detect errors and guide repair and with the necessary skills of the worker to form an integrated support system.

Another highly significant and often discussed context of application for support technologies are nursing homes and geriatric care. *Jannis Hergesell* and *Arne Maibaum* have observed how the introduction of an assistance systems for caregivers is accompanied by interests and differing interpretations of various stakeholders that do not simply match with the demands of elderly care. Additionally, unintended effects must be recognized and managed in interaction processes if the supposed support technology is to function flawlessly. Their contribution presents also a meticulous account of a dynamic field of expectation already alluded to in Part I of the book.

Construction sites are, to a certain extent, structured environments in which craftsmen can profit from augmentation supporting their work and performance. *Kathrin Nuelle, Sabrina Bringeland, Svenja Tappe, Barbara Deml*, and *Tobias Ortmaier* present work on a particular head-mounted solution that enables craftsmen to improve their drilling accuracy and to get a 3D image of the drillhole pattern as a whole. The authors describe the system setup and how the real-world positions can be transformed into the display. Finally, they evaluate their results in a study with professional users.

It is commonly ignored or disregarded that support technology is mainly introduced and implemented in organizational contexts. Nursing homes, industrial plants, or craft enterprises differ in their purpose, but all have to deal with organizational issues. *Daniel Houben, Annika Fohn, Mario Löhrer, Andrea Altepost, Arash Rezaey*, and *Yves-Simon Gloy* introduce a comprehensive heuristic that enables developers of technology to assess how assistance/support systems affect organizational processes. Two of the main conceptual tools that allow such an analysis are the management of uncertainty and negotiated orders in organizations.

Support technology can also be used to develop other support technologies. *Gabriele Bleser, Bertram Taetz*, and *Paul Lukowicz* do not consider a particular context or domain of application but rather a form of deployment. Their area of expertise is wearable sensor networks that are deployed to collect reliable in-field data of activities and body movements. With painstaking attention to technical detail, they present two forms of deploying wearable sensor networks: one for gathering information about joint kinematics and one for activity recognition.

Agostino Stilli, Kaspar Althoefer, and *Helge A. Wurdemann* conclude this third section of the book. Their contribution is concerned with soft robotics which is generally considered as the course robotics should take to realize technical support that people really want. Softening robots not only by programming but also by utilizing soft materials, joints, and actuator principles allow robots to come closer to humans in a safe way and to mimic the complex and smooth movements of organic bodies. The authors tackle one of the main challenges of this research area: the distinction between softness and stiffness. By getting biological inspiration from the arms of an octopus, they work on the idea of fabric-based antagonistic stiffening mechanisms and show how soft collaborating robots might be thus designed to handle the soft/stiff distinction in a dynamic way.

Support Technologies for Industrial Production

**Robert Weidner, Bernward Otten, Andreas Argubi-Wollesen
and Zhejun Yao**

Abstract In industrial production, the construction industry, and in daily life, manual activities will remain of central importance. The manifold environments and tasks require the development of different support systems. The choice of a system for a concrete application strongly depends on the context. In addition to (semi-) automated solutions, wearable systems for physical support have been developed in recent years. This chapter introduces needs, context-relevant requirements, and exemplary systems for support in industrial production.

1 Introduction

Today, and also in future, human beings will continue to play an important role in value chains—especially for tasks with high variety or complexity. The reasons are, on the one hand, the unique intelligence and skills of humans, and on the other hand, the high financial and technical costs of (semi-) automated solutions. Corresponding tasks, such as the handling of tools and components, possibly used in aggravating and non-ergonomic positions (e.g., at or above head level), can lead to high physical loads. As a result, disorders and injuries of the musculoskeletal system might occur.

R. Weidner (✉) · B. Otten · A. Argubi-Wollesen · Z. Yao
Laboratory of Manufacturing Technology, Helmut Schmidt University/University of the Federal
Armed Forces Hamburg, Holstenhofweg 85, 22043 Hamburg, Germany
e-mail: robert.weidner@hsu-hh.de

B. Otten
e-mail: ben.otten@hsu-hh.de

A. Argubi-Wollesen
e-mail: argubi-wollesen@hsu-hh.de

Z. Yao
e-mail: zhejun.yao@hsu-hh.de

R. Weidner
Chair of Production Technology, University of Innsbruck, Innsbruck, Austria

© Springer Nature Switzerland AG 2018   149
A. Karafillidis and R. Weidner (eds.), *Developing Support Technologies*, Biosystems &
Biorobotics 23, https://doi.org/10.1007/978-3-030-01836-8_14

In addition, challenges due to demographic change, such as the longer working lives and the skills shortage, also demand for novel solutions to support the employees.

The just described initial situation defines starting points for a direct co-operation of humans and technology in form of a wearable, technical system for physical support or physical-cognitive relief. Correspondingly, hybrid system approaches, e.g., the concept of Human Hybrid Robot (HHR) [Wei13a, Wei13b], are proposed in order to be able to use the complementary and partly contrasted skills and abilities of humans and technology and to synchronize them in time and space. This includes, on the one hand, the cognition, sensory flexibility, and learning ability of the human being, and on the other hand, the endurance, repeatability, and reliability of technical systems.

In the following sections, typical activities in industrial production and the corresponding support demands are described. Then, context-relevant requirements are identified regarding the users and the activities. Based on the requirements and the Human Hybrid Robot (HHR) approach some wearable support systems for different contexts are presented. Finally, the potential of those support systems is discussed.

2 Typical Activities

In all industrial sectors working contexts can be distinguished based on the movement patterns of the activities. Although the working contexts differ in their specific characteristics (e.g., loads to be moved, speed, and time), they represent comparable physical actions. Widely used tools for risk assessment at the workplace, such as the Key Indicator Method (KIM) [Ste07] or the OVAKO Working posture Analyzing System (OWAS) [Bra17], provide an overview of the criteria by which manual handling of loads can be differentiated. These activities are too numerous to be completely enumerated here. A selection of the most common activities and the stressed body parts in those activities are summarized in Table 1. Specific and clearly distinguishable activities that occur repeatedly in manual production are, e.g., lifting and carrying, pushing and pulling, and working at or above head level. Further activities are bending the knee, bending the back, and kneeling. Work-specific activities can consist of some of these activities as well as any combination of them.

3 Context-Relevant Requirements

Although wearable support systems (i.e., exoskeletons) tend to be universal technical solutions, they have to be adapted individually to the context—especially to the users and activities. The relevant requirements can differ significantly. Primarily, the support system must allow all necessary movements of the user. This can be achieved by proper coupling of structural elements over discrete degrees of freedom, e.g., [Wei14, Ott16, Mey16]. In addition, it is possible to match the technical system to the user's motion via soft structural elements, e.g., [Wei16, Yao17].

Table 1. Overview of stressed body parts in the most common activities regarding the demands on strength, movement dynamics, and range of motion

		Stressful body parts									
		Wrist	Arm	Neck	Shoulder	Back	Leg				
Activities	Lifting and carrying	◕	◕	◔	●	●	◕		Strength	Movement dynamics	Range of motion
	Working at or above head level	◑	◐	◔	●	◐	○				
	Pushing and pulling	◔	◔	○	◔	●	◑				
	Drilling and screwing	◕	◐	◔	●	◐	○				

Legend ○ low ... ● high

Figure 1 summarizes the key requirements on exoskeletal systems in the overall context.

Wearable systems for physical support usually have a serially coupled robotic structure, which transfers excessive load from the stressed body parts to somewhere else or supplies an external supporting force to the overloaded body parts. The serial structure can be designed in two different ways: a biomechanically equivalent structure or an "end-effector" structure. In the case of a biomechanically equivalent structure, the attempt is to exactly mirror the kinematics of human motion and the human's ranges of motion for each degree of freedom, e.g., the position of the exoskeletal arm follows the position of the human arm [Ott16]. In the other case, the "end-effector" structure has a non-anthropomorphic architecture. Only the physical interfaces, which connect to the user's body, follow exactly the position of the user's body [Wei14]. Gradations between these two ways of serial coupling are also possible, e.g., [Mey16].

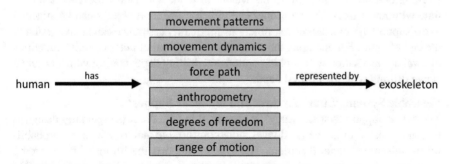

Fig. 1 Key requirements on exoskeletal systems

It is difficult to map all degrees of freedom of the human skeletal structure, especially at complex joints, like the shoulder joint whose rotation axes shift during the movements. Moreover, the exoskeletal structure should be highly adjustable for different users due to the deviation of the different users' bodies. This adjustability and flexibility can be achieved by appropriate internal architecture or the selection of suitable material.

Table 1 shows that not all body parts have the same stress level in each activity. Even the same body part does not need the same level of support for different activities. In order to reduce the overloads on the targeted body parts, wearable systems with different supporting foci are needed. Moreover, depending on the strength, movement dynamics, and range of motion required by the activity, different system properties are needed. For instance, the working load determines the strength of the support force and the movement dynamics required by the task determines the system dynamics. In addition, the working duration and environment influences the choice of the system's power supply.

4   Exemplary Solutions for Different Contexts

Based on the described approach and the requirements mentioned above, this section presents several exemplary support systems for manual tasks in industrial production. These systems aim to reduce the overload of the employees and to improve the ergonomic conditions at the workplace. They cover the most common manual activities in industrial production and address the diverse stressed body parts.

Wearable Tool Holding System "Jonny"
Figure 2a shows the support system "Jonny", a passive support system designed for handling heavy tools and components [Wei14]. It has an "end-effector" structure with a tool attachment based on a Steady Cam mount and is exemplarily equipped with a drilling end-effector. The structure distributes the weight of the drilling machine to the torso and releases the upper extremities from overload. This improves the working ergonomics and allows the wearer to work with heavy tools for a longer time without distress. With a proper end-effector, different objects can be attached to the support system. Moreover, functions for quality assurance can be integrated in the end-effector. For example, the presented drilling end-effector contains functions of level compensation with locking option and drilling depth control which ensure a stable drilling with the desired depth.

Wearable System "Lucy" for Arm and Shoulder Support
The active support system "Lucy" is presented in Fig. 2b. It is especially designed to make tasks at or above head level more comfortable and to reduce the probability of musculo-skeletal disorders. The system supports the lifting of the wearer's arms against gravity by using pneumatic actuators [Ott16]. Thanks to the flexible back structure and the biomechanically equivalent design of the system joints at the shoulder, the user can move freely without discomfort. In order to realize an intu-

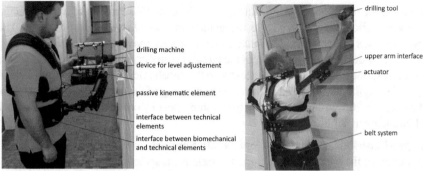

(a) support system "Jonny" for handling tools

(b) Support system "Lucy" for tasks at or above head level

(c) Exoskeletal spine module

d) Support system with paper-lamella elements

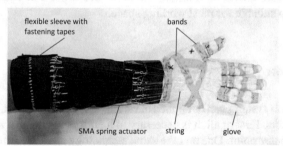

(e) Muscle glove for support of gripping

Fig. 2 Exemplary support systems for different contexts

itive operation and a better user experience, the support force is regulated during the motion and can be individually adapted to users and tasks.

Exoskeletal Spin for Back Support

Figure 2c shows one module of an exoskeletal spine, which aims to reduce stress and strain on the user's back during manual tasks. The module consists of two connected elements, attempting to replicate the connection of two vertebrae [Mey16]. The two elements are connected in such a way that they move against each other on a common circular path. The center of the circular path matches the rotational center of the corresponding human vertebrae. Due to this mechanical behavior, the exoskeletal spine is lengthened and thus follows the lengthening of the user's back. The length of the exoskeletal spin can be adjusted to different users by connecting a certain number of elements in series.

Paper-Lamella Elements for Upper Body Support
A variant of the paper-lamella support system is presented in Fig. 2d. It is a garment with integrated paper-lamella elements that consist of a chamber and several overlapping paper sheets inside the chamber. The stiffness and tension of the element vary when the pressure inside the chamber changes. In this way, the paper-lamella support technology can help to stabilize working posture. For different applications, the element exists in three variants: with constant length, with flexible length, and with a rotational degree of freedom. The elements can also be connected to each other in various arrangements.

Muscle Glove for Hand and Wrist Support
The muscle glove illustrated in Fig. 2e aims to support people doing hand-intensive activities by reducing muscle contraction force and time in a defined work-rest cycle [Yao17]. Based on a biomimetic design it attempts to replicate the salient features and functionalities of the human hand in great detail. The muscle glove has a textile-based soft structure which contains string, band, and a shape memory alloy (SMA) coil actuator as artificial tendon, pulley, and muscle. Thanks to the soft and biomimetic structure, it is inherently compatible with the human kinematics and can perform natural hand movements. In addition, the miniature SMA coil actuator allows a compact design suitable for most working space.

5 Discussion

Wearable support systems can be applied in various contexts to support, assist, or help industrial workers. First of all, it is necessary to identify the effective and, above all, goal-oriented entry point. One and the same technology can have different degrees of impact in different application contexts, i.e., the same support system may be very effective in one context but remain ineffective in another. The influence factors contingent on the choice of a context-relevant support system are illustrated in Fig. 3.

The use of support technology is mainly driven by organizations (e.g., manufacturing companies and associations) or by employees for different purposes. These purposes could be, for example, improving production quality and productivity as well as improving ergonomic conditions. In order to improve the support systems and their matching to the working context, a profound understanding of support situations is necessary.

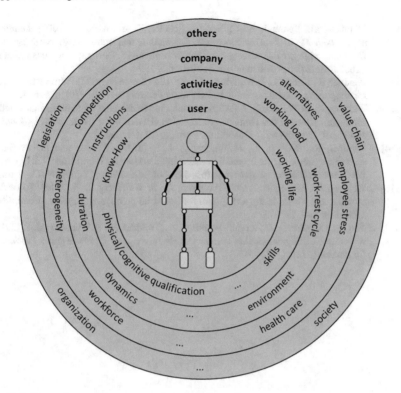

Fig. 3 Influence factors for a context-relevant choice of support system

Acknowledgements This research is part of the project "smartASSIST—Smart, AdjuStable, Soft and Intelligent Support Technologies" funded by the German Federal Ministry of Education and Research (BMBF, funding no. 16SV7114) and supervised by VDI/VDE Innovation + Technik GmbH.

References

[Bra17] Brandl, C., Mertens, A., & Schlick, C. M. (2017). Effect of sampling interval on the reliability of ergonomic analysis using the Ovako working posture analysing system (OWAS). *International Journal of Industrial Ergonomics, 57,* 68–73.

[Mey16] Meyer, T., & Weidner, R. (2016). Exoskelettale Wirbelsäulenstruktur zur Aufnahme und Umleitung von Kräften zur Rückenentlastung. In R. Weidner (Ed.), *Proceedings of the 2nd transdiciplinary conference "Technical Support System, that poeple really want"* (pp. 567–576).

[Ott16] Otten, B., Weidner, R., & Linnenberg, C. (2016). Leichtgewichtige und inhärent biomechanisch kompatible Unterstützungssysteme für Tätigkeiten in und über Kopfhöhe. In Weidner, R. (Ed.), *Proceedings of the 2nd transdiciplinary conference "Technical Support System, that poeple really want"* (pp. 495–505).

[Ste07]   Steinberg, U., Behrendt, S., Caffier, G., Schultz, K., & Jakob, M. (2007). *Leitmerk-malmethode Manuelle Arbeitsprozesse. Erarbeitung und Anwendungserprobung einer Handlungshilfe zur Beurteilung der Arbeitsbedingungen*. Dortmund: Bundesanstalt für Arbeitsschutz und Arbeitsmedizin, Projektnr. F, 1994.

[Wei13a]   Weidner, R., Kong, N., & Wulfsberg, J. P. (2013). Human Hybrid Robot: A new concept for supporting manual assembly tasks. *Production Engineering, 7*(6), 675–684.

[Wei13b]   Weidner, R., & Wulfsberg, J. P. (2013). Mensch-Maschinenhybride in der industriellen Montage - Konzept des Human Hybrid Robot (HHR). In *wt Werkstattstechnik online 103* No. 9, (pp. 656-661). Düsseldorf: Springer-VDI-Verlag.

[Wei14]   Weidner, R., Redlich, T., & Wulfsberg, J. P. (2014). Passive und aktive Unter-stützungssysteme für die Produktion - Konzept des Human Hybrid Robot (HHR). In *wt Werkstattstechnik online 104* Nr. 9, (pp. 651–666). Düsseldorf: Springer-VDI-Verlag.

[Wei16]   Weidner, R., Meyer, T., Argubi-Wollesen, A., & Wulfsberg, J. P. (2016). Towards a modular and wearable support system for industrial production. *Applied Mechanics & Materials, 840,* 123–131.

[Yao17]   Yao, Z., Linnenberg, C., Argubi-Wollesen, A., Weidner, R., & Wulfsberg, J. P. (2017). Bio-mimetic design of an ultra-compact and light-weight soft muscle glove. *Production Engineering, 11*(6), 731–743.

Assistance Systems for Production Machines in the Textile Industry

Yves-Simon Gloy

Abstract Assistance systems for production machines have to fulfill several functions depending on specific tasks and users. The overall purpose of the use of an assistance system in production is to raise the efficiency of the production. In this text, one approach for an assistance system for a weaving machine is presented. Based on a systematic approach, a system is designed in order to facilitate the repair of broken weft yarns. Using this approach an assistance system with a high general user acceptance can be achieved. Furthermore, the use of assistance systems will have an impact on the necessary qualification of operators in the weaving mills.

1 Motivation

Various approaches are being developed in the context of Industry 4.0. When looking at jobs in production in the wake of this development, so-called assistance systems become more and more important. Such assistance and support systems for production machines should realize physical relief for an operator, provide guidance on how to interact with a production machine, simplify the documentation, allow learning during the process of work, protect workers against accidents, provide possibilities for exchange between employees, adapt to the needs and previous knowledge of the user, and increase the efficiency of a production machine (see Fig. 1).

With regard to the interaction with a production machine, an assistance system is expected to support the user during maintenance, inspection and service, learning, documentation, equipment of a machine, operation, optimization of, e.g., production efficiency and energy consumption, as well as repair processes.

Y.-S. Gloy (✉)
Institute for Textile Technologies, RWTH Aachen University, Otto-Blumenthal-Str.1, 52074 Aachen, Germany
e-mail: yves.gloy@ita.rwth-aachen.de

© Springer Nature Switzerland AG 2018
A. Karafillidis and R. Weidner (eds.), *Developing Support Technologies*, Biosystems & Biorobotics 23, https://doi.org/10.1007/978-3-030-01836-8_15

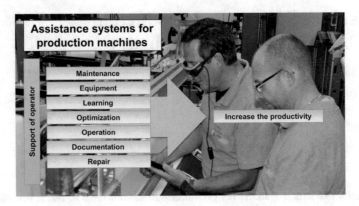

Fig. 1 Assistance systems for production machines

From an engineering point of view, the ultimate goal of such assistance systems is to increase the productivity of the production by extending the main usage time and reducing downtime. In order to realize this, the system must always be adapted to the user and the application.

In what follows, a realized assistance system for a weaving mill is presented.

2 Application

The presented assistance system supports the elimination of a weft breakage. This task has the following steps:

1. Remove remaining weft threads from the channel and reed.
2. Go to the creel and check the status of the spool.
3. When the bobbin is empty, replace the bobbin with a new one with yarn of the same type.
4. If the bobbin is not empty, guide the yarn through the brake on the gate.
5. Feed the yarn to the feeder, press the yellow button there and use the compressed air to feed the yarn into the feeder.
6. Activate the feeder by moving the yellow button on the side to collect sufficient length of yarn on the drum. Make sure that the feeder remains switched on.
7. Take the insertion hook, which is located on the loom.
8. Guide the insertion hook through the balloon breaker.
9. Attach the yarn to the insertion hook and pull it through the balloon breaker.
10. Return the insertion hook.
11. Activate Manual Mode on the loom's touch screen.
12. Select in the menu the corresponding channel of the weft insertion.

13. Lead the thread to the beginning of the first main nozzle, activate the nozzle at the terminal of the weaving machine and guide the thread into the nozzle with the help of compressed air.
14. Lead the thread to the beginning of the second nozzle, activate the nozzle at the terminal of the weaving machine and guide the thread into the nozzle with the help of compressed air.
15. Remove possible thread remnants from the weft channel.

The assistance functionalities are integrated into the weaving machine or other additional devices. The assistance system uses elements for monitoring and communication.

2.1 Monitoring System

The monitoring system is developed to detect the state of the weft insertion on an OmniPlus 800 weaving machine of the company Picanol n.v., Ieper, Belgium. The weaving machine is already equipped with a monitoring system, which is able to detect weft defects and interrupt the weaving process. It is completely integrated into the weaving machine and does not disturb its error-free run. Furthermore, it displays error messages about weft breakage via a touch screen and stores information about it in the logbook of the machine.

The integrated monitoring system of the loom fulfills all previously mentioned functional requirements. In addition, even the non-functional requirements in terms of speed, accuracy, and reliability are met. Therefore, this monitoring system is used in the further development.

2.2 Communication System

The communication system is responsible for the data transfer between the integrated monitoring system and the assistance system. In the current weaving machine configuration, bidirectial communication between the weaving machine and a soft programmable logic controller was realized (PLC) [Sag14, San12]. The loom is digitally controlled (machine parameters can be stored in an internal memory). Therefore, a PLC was used to read this data with an external computer via TCP/IP interface.

The communication between loom and external computer was realized by means of hardware and software of the company iba AG, Fürth, that is, a modular system for the acquisition of measurement signals and a software for the processing of these signals. Together, they form a PLC to enable bidirectional communication. Thus, the loom parameters can be deciphered and changed on an external computer.

Furthermore, an Object Linking and Embedding for Process Control (OPC) server has been implemented to provide remote access to the machine. OPC is a software interface standard that allows Windows programs to communicate with industrial devices. By using an OPC server, mobile devices can access the loom, e.g., via a web-based interface. Within the scope of this work the software Atvise of the company CERTEC EDV GmbH, Eisenstadt, Austria is used to provide a real-time and online display for a range of machine parameters. This can be addressed and modified directly from mobile devices by taking the IP protocol of the browser of the mobile device. In this form, the design of the communication system meets also all pre-defined requirements.

2.3 Assistance System

The monitoring system recognizes the status of a critical event followed by an interpretation. The interpretation is done by the machine. The machine has stored meanings for signals in the data memory. If no error occurs, the web process continues. Otherwise, an error is communicated to the weaving machine, which interrupts the weaving process and contacts the assistance system.

The monitoring system communicates directly with the weaving machine and exchanges information about the current status of the critical event. The weaving machine communicates the status of the critical event to the assistance system and to the operator via the human-machine interface of the weaving machine (touch screens, signal lights). The assistance system provides information to the operator via a mobile interface along with a series of hints based on augmented reality (AR). The operator interacts with the assistance system via a menu and gives his consent to start a repair process, see Figs. 2 and 3.

The programming is carried out in the Augmented Reality Experience Language (AREL) of the company Metaio GmbH, Munich. Thus, a cross-platform software for Android, iOS and Windows devices can be created. Platform independence is achieved using standard web technologies such as HTML5, Extensible Markup Language (XML) and JavaScript. Therefore, programming in AREL shows similarities to webdesign: Static XML files define the content and settings of the tracking; JavaScript makes it possible to visualize AR scenarios in a web browser and HTML and Cascading Style Sheet (CSS) allows display in web browsers.

Essentially, the elements text, 2D, and 3D content are realized. Text is mainly used to communicate with the user and provides instructions for repair. The text is created using HTML programming and adapted for aesthetic reasons with CSS commands.

2D images are needed in the interface layout or in the augmentation of the user's environment. Arrows, e.g., can be displayed in the interface to show the location of the next machine element, or a label is projected onto elements so that they are recognized more quickly by the user.

Fig. 2 AR based assistance system for weaving machines

Image to indicate camera
perspective for finding 3D map

Instruction to find trackable

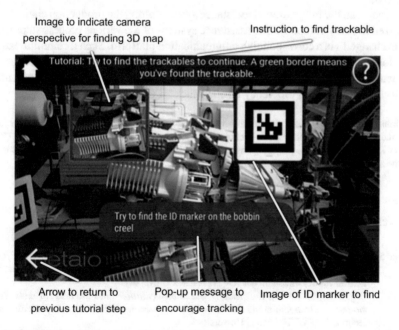

Arrow to return to
previous tutorial step

Pop-up message to
encourage tracking

Image of ID marker to find

Fig. 3 Tutorial view for detection of trackables [Lon15]

3D content such as CAD models or 3D arrows or machine elements are used to augment the user's environment. To generate such 3D content, the software Autodesk Inventor of the company Autodesk Inc., San Rafael, USA is used together with the tool Blender of the company Blender Foundation NPO, Amsterdam, The Netherlands.

Figure 3 show the assistance for the elimination of a weft defect. The user is guided by the necessary repair process on the loom. For this purpose, he receives appropriate graphical information as to which work steps have to be performed and in what sequence, so that the defect is remedied.

3   Discussion

The first effects of the use of such assistance systems can already be recognized. Since the work content will change in part, more and more staff skills will be needed. A learning-friendly design of the assistance systems would support the required competence development. When employees internalize the process and machine technology the risk of production stoppages and accidents at work will be reduced [Wis18].

In general, the integration of assistance systems along the production chain can be realized in every company. The assistance systems of the various process steps could be exchanged via a corresponding communication platform. Such a communications platform could also make it possible to establish communication between assistance systems of various companies. The goal should be to find and use further potentials to increase production efficiency through such an exchange of information.

Acknowledgements I thank the Federal Ministry of Education and Research (BMBF) for funding the research group "New Socio-technical Systems in the Textile Industry (SozioTex)" (FKZ: 16SV7113), as well as the project sponsor VDI/VDE Innovation + Technik GmbH for the support in the application and execution of the project.

References

[Lon15]   Longé, G. (2015). *Assistance systems for industry 4.0 in weaving machinery with special focus on augmented reality-based applications for smart glasses*. Masterarbeit. Aachen: Institut für Textiltechnik der RWTH Aachen University.

[Sag14]   Saggiomo, M. (2014). *Entwicklung und Implementation eines Modells zur mehrdimensionalen Selbstoptimierung des Webprozesses; Masterarbeit*. Aachen: Institut für Textiltechnik der RWTH Aachen University.

[San12]   Sandjaja, F. (2012). *Implementierung eines Konzeptes zur Selbstoptimierung einer Luftdüsenwebmaschine; Diplomarbeit*. Aachen: Institut für Textiltechnik der RWTH Aachen University.

[Wis18]   Wischmann, S., & Hartmann, A. (2018). *Zukunft der Arbeit – Eine praxisnahe Betrachtung*. Berlin, Heidelberg: Springer.

Interests and Side Effects
in the Technicization of Geriatric Care

Jannis Hergesell and Arne Maibaum

Abstract Technical assistance is currently being promoted in the field of elderly care as a means of coping with the challenges of demographic change. Nursing is one area where such technicization is taking place. The implementation of assistive technologies in socially complex areas, such as nursing, requires integrating a wide range of actors and their interests to make full use of technological potential. Apart from their intended use, assistive technologies also produce non-intended effects. The design of innovative technologies must take this into account to prevent the technology from failing because of resistance and incompatibilities. This article uses nursing assistance as an example of technical assistance systems for geriatric care to discuss the social dimensions in relation to the desired technicization of nursing care and the fundamental conflicts that characterise technology development and implementation.

1 Needs, Potentials, and Problems of Assistive Technologies for Geriatric Care

Unlike any other part of society, technical assistance is currently being promoted in the elderly care sector. The driving force behind this is the challenge of demographic change, which threatens the care of the elderly in two ways. First, a high increase in the need for care is to be expected [Hir14]. Second, shrinking working-age population suggests a decline in nursing staff, less care by family members, and fewer contributors to social security systems [BPB11, Now13]. The 'nursing crisis', which is being discussed today, poses a threat to an affordable, qualified, and ethically acceptable level of care provision in the future.

J. Hergesell (✉) · A. Maibaum
Department of Sociology, DFG Graduate School "Innovation Society Today", Technical
University of Berlin, Berlin, Germany
e-mail: jannis.hergesell@innovation.tu-berlin.de

A. Maibaum
e-mail: arne.maibaum@innovation.tu-berlin.de

© Springer Nature Switzerland AG 2018   163
A. Karafillidis and R. Weidner (eds.), *Developing Support Technologies*, Biosystems &
Biorobotics 23, https://doi.org/10.1007/978-3-030-01836-8_16

Previous attempts to solve these problems, such as reforms of care insurance or recruitment campaigns for the nursing profession, have not shown the hoped-for effect so far. They have been replaced by a new strategy to solve the problems of elderly care: promotion of the development and implementation of technical assistance (see [Alb15, pp. 347, Nie16, p. 27]). These sensor-based assistive technologies are intended to increase the quality of life for the care receivers by supporting their independence and creating security. The care givers should be relieved and supported by the assistive technologies. In addition, the implementation of assistive technologies also promises the optimization of work organization and a general increase in the efficiency of care work [Hie15, Buh15, Küh15]. In short, innovative assistive technologies address all the current problems of geriatric care and provide a promising solution strategy for the future design of elderly under the conditions of demographic change. Particularly political actors recognize the potential of these technologies and actively promote them in research programs.

Despite these promises, the actual use of assistive technologies has so far been relatively low. Often their use is even resisted. Only a few assistive technologies have evolved from the phase of development and testing or are 'market-ready' [Wei15, End16]. Care receivers and nursing staff often disapprove of new technologies. They fear the 'dehumanization' of care. The question arises as to why there is such a great discrepancy between potential and actual use of assistive technologies.

2   Conflicting Patterns of Interpretation and Interests in Everyday Care

On closer examination, it becomes clear that apparently shared objectives connected with technicization are ambiguous or even contradictory. In consequence, there is a lack of integration between the different logics of the actors involved.

In order to understand why there is a difference between the potential of assistive technologies and the actual application in geriatric care, the perspectives of all actors must be understood. For a better understanding of the complex social dimensions involved in technologically assisted care, the knowledge and power relationships of all the actors involved and the inscriptions of these into the technologies must be fundamentally disclosed. This approach makes it possible to draw conclusions about the social structures of elderly care, thus explaining the scant use of assistive technologies to date.

Empirically, six relevant groups of actors can be identified: (1) nursing managers; (2) nursing staff; (3) care receivers; (4) relatives and social environment; (5) technology developers and providers, as well as (6) social and health policy actors and funders.

Various patterns of interpretation and perception can be assigned to these groups, each with a distinct approach to the problems, which are to be solved by the technology. For the analysis, the patterns can be divided into two categories:

Category one consists of the developers and providers of care technology (group 5), as well as the social and health policy actors and funders (group 6). This group has a pattern of interpretation based on economic-instrumental criteria. They act in the logic of increasing efficiency in the sense of a cost-benefit ratio (instrumentally rational).

Category two comprises the professional nurses (group 2), nursing managers (group 1), care receivers (group 3), and family members (group 4). They have internalized patterns of perception and interpretation guided by nurturing values. Nurturing values are oriented towards the greatest possible preservation of the autonomy of the care receivers, the individual care, and professional nursing competence.

These patterns seem to have generally shared, problem-equivalent perceptions of problems, for example, the constant lack of personal nursing and time resources. Nevertheless, the different logics draw on completely different conceptions of problem-solving strategies. In addition, these different groups have unequal chances to impose their patterns on each other, which makes an understanding even more difficult.

3   Care Immanent and Economic Logics

Immanent nursing patterns of perception and interpretation include a humanistic understanding of care, which provides technically correct medical as well as qualified psychosocial care. From this perspective, everyday care should be designed towards the individual needs of the care receivers. This care should go beyond the mere basic care and guarantee quality of life in the sense of respecting the dignity of the care receivers to maintain autonomy and offer participation in society. The conditions for fulfilling these objectives are the deployment of highly qualified nurses and a work organization that provides sufficient time resources.

In contrast, economic perceptions and interpretive patterns focus on the financial feasibility, legal aspects, such as reimbursement of costs or compliance with official standards, and effective work organization. The focus is less on the everyday care situations, or the perception of the individual case, but rather on the economic handling of resources in the care sector. These goals are to be achieved by an increase in efficiency, that is, the most efficient use of the available resources and cost reduction by avoiding care work deemed unnecessary.

4   Consequences and Side Effects of Failed Technology Development

Different kinds of knowledge are reflected in the funding, development, and implementation of technological assistance. Often, an apparent semantic match of technicization goals is merely a lexical one. Although the same vocabulary is used, the relevant actors refer to fundamentally different meanings. For example, nurses seek to preserve self-determination of the care receivers by focusing on psychosocial care measures and aiding the care receivers in individual necessity fulfilment. To do so, they expect technology to dispense them from non-core care tasks such as filing or repetitive control chores.

In contrast, technology developers and political actors understand the benefits from assistive technologies as the least disruptive possible intervention from the caregivers in the lives of the care receivers. Autonomy in this logic should be guaranteed, preferably in ambulatory care, by controlling the physical security, which substitutes physical presence of caregiver.

So, even if the carriers of the different knowledge stocks lexically use the same notion of self-determination, they nevertheless aspire to very different objectives of the technology application. Subsequently, since the funding, development, and implementation is dominated by the technology developers and political actors, the economic interpretation persists during the process, which results in inadvertently writing their conceptions care into the assistance systems. Due to these rather invisible power balances, it has not been possible to develop an assistive technology, which corresponds to the interests of the caregivers and the care receivers despite the effort of an integration of the positions.

The failing integration of the different interests can be illustrated by the empirical example of an assistant during the night-watch in a nursing home for people living with dementia. The typical problematic situation is that often only one nurse is responsible for the well-being of more than twenty care receivers. Due to the dementia induced shifted day-night rhythm of the residents, their status is often hard to differentiate. To support the nurse, a sensor-based assistance system is introduced. The assistance system uses motion detectors to alarm the nurse when a resident has stood up or left the room. The intention is to guarantee an efficient protection against falls, as well as to avoid disturbances of the inhabitants caused by nocturnal inspections of the nurse. The nurse is relieved from insecurity and repetitive controls.

Unlike the projected outcome, however, it can be observed that the nurses continue to control the rooms on an hourly basis. On the one hand, they perceive a high responsibility for the care receivers and are not willing to cede this to an assistance system. On the other hand, and more significantly, the nurses' supervision is not just limited to mere safety aspects. Controlling the care receivers is part of psychosocial care especially to those with a shifted rhythm. It becomes an important task during the night-watches. Consequently, the care receivers do not interpret the nurses' controls as a disturbance or as an infringement of their autonomy. Instead it is an integral part of the nursing measure.

The result is that the actual goal of the assistance system, despite being designed in accordance to enhance the self-determination by reducing the control cycles, is neither achieved nor desired. The inherent concept inscribed into the technology, the idea of relief, does not correspond with the care plans and the anticipated needs of the nurses. Despite the lexical similarity we see a discord between the economic and the nursing logic.

This mutual limitation of economic and nursing logics is a structural characteristic which has existed since the foundation of the nursing profession. Since this beginning there has been a conflict between the economic and professional-ethical interests in care resulting from a separation in the management and the everyday care work, carried out by separate groups of people. Central problems regarding care, such as the need for staff, the low social gratification of the nurses, the dominance of scientific medicine against care, and the delayed professionalization are due to this separation of the different logics.

The beginning of the technicization of nursing now runs the risk of continuing this process and thus also consolidating the traditional misinterpretation and communication between the care-givers and the political actors. The basic discrepancy of the different patterns of perception and interpretation, which are internalized by different groups of actors during their (occupational) socialization, cannot be overcome by dialogical strategies only. The risk of misinterpretation between the different cultures remains persistent even when moderated by third parties, for example in participatory technology development. Even if individual concepts such as increasing efficiency and quality of life seem to be consensual, this is true only on a lexical, but not on a semantic level.

5   Integration of Knowledge by Participatory Technology Development

Assistive technologies have implications for the social structures of care beyond their intended benefits. These side effects are often latent and can only be demonstrated empirically by long-term studies. Even today, however, development tendencies can be seen. It can be observed that the technologies used so far are those that draw on an increase of efficiency. Increase in efficiency relies on standardization and selection of reasonable and not reasonable care measures, whereas reasonable often means economically reasonable. In consequence, technologies are unintentionally produced to rely less on traditional, care-intrinsic knowledge, like, for example, the psychosocial aspect of night-watch control visits to dementia patients. In the long term this could lead to a deprofessionalization of the nurses. Nursing competence might be replaced by standard operational procedures during the introduction of the technology.

In order to reduce the great discrepancy between the potential of assistive technology systems and their currently limited use in nursing homes in the future, the different interests of the involved actors have to be recognized and integrated. This requires an informed reconstruction of the field's evolved actor constellation as well as careful attention to the implicit balance of power within this actor constellation. Only in this way is it possible to recognize the latent power structures and to counteract them when necessary. In the case of nursing care, this might possibly mean a stronger involvement of the nursing staff in the development process. This integration, however, has to go beyond a purely dialogical participation process. For example, nursing staff could be involved in the planning stage of technology development. In this way, a political emancipation of the nurses on the administrative level seems possible and might allow the development of integrated assistive technology systems.

References

[Alb15]   Albrecht, M., Hinding, B., & Kastner, M. (2015). Das Innovationspotential in Pflege- und Sozialberufen. In Jeschke, S., Richert, A., Hees, F., & Jooß, C. (Eds.), *Exploring demographics. Transdisziplinäre Perspektiven zur Innovationsfähigkeit im demografischen Wandel* (pp. 347-359), Wiesbaden: Springer.

[BPB11]   Bundeszentrale für politische Bildung (2011): Demografischer Wandel in Deutschland. Bevölkerungsentwicklung: Soziale Auswirkungen. http://www.bpb.de/politik/innenpolitik/demografischer-wandel/75997/soziale-auswirkungen?p=all, cited June 5, 2017.

[Buh15]   Buhr, D., Haug L., & Heine, T. (2015). Pflegeassistenzen. In Weidner, R., Redlich, T., & Wulfsberg, J. (Eds.), *Technische Unterstützungssysteme* (pp. 200–202). Wiesbaden: Springer.

[End16]   Endter, C. (2016). Skripting age—the negotiations of age an aging in ambient assisted living. In Domínguez-Rué E., & Nierling, L. (Ed.), *Ageing and technology. Perspectives from the social sciences* (pp. 121–141). Bielefeld: Transcript.

[Hie15]   Hielscher, V., Nock, L., & Kirchen-Peters, S. (2015). *Technikeinsatz in der Altenpflege. Potentiale und Probleme in empirischer Perspektive.* Baden-Baden: Nomos.

[Hir14]   Hirschberg, K. (2014). Einflüsse demografischer Veränderungsprozesse auf die Arbeitssituation der Altenpflegekräfte – Faktenlage. In Behr, T. (Ed.), *Aufbruch Pflege. Hintergründe – Analysen – Entwicklungsperspektiven* (pp. 193–199). Wiesbaden: Springer-Gabler.

[Küh15]   Kühne, H. (2015). Chancen und Herausforderungen. Nutzerbedarfe und Technikakzeptanz im Alter. Technikfolgenabschätzung – Theorie und. *Praxis, 24*(2), 28–35.

[Nie16]   Nierling, L., & Domínguez-Rué, E. (2016). All that glitters is not silver—Technology for the elderly in context. Introduction. In Domínguez-Rué, E., & Nierling, L. (Ed.), *Ageing and technology. Perspectives from the social sciences* (pp. 9–27). Bielefeld: Transcript.

[Now13]   Nowossadeck, S. (2013). *Demografischer Wandel, Pflegebedürftige und der künftige Bedarf an Pflegekräften. Eine Übersicht* (Vol. 56, pp. 1040–1047). Bundesgesundheitsblatt – Gesundheitsforschung – Gesundheitsschutz.

[Wei15]   Weinberger, N., & Decker, M. (2015). Technische Unterstützung für Menschen mit Demenz? Zur Notwendigkeit einer bedarfsorientierten Technikentwicklung. *Technikfolgenabschätzung – Theorie und Praxis, 24*(2), 36–45.

Mobile Augmented Reality System for Craftsmen

Kathrin Nuelle, Sabrina Bringeland, Svenja Tappe, Barbara Deml and Tobias Ortmaier

Abstract A mobile augmented reality system for visual guidance of craftsmen is presented. The integrated pico-projector displays constructional information within the workers field of vision. It displays borehole locations in correct world pose and helps workers aligning a drill perpendicular to a wall. To create assisting images using coordinate transformation, rtab-SLAM and camera-projector calibration determine the relative pose between the environment and the projector. To evaluate the systems usability, a user study was performed on drill aligning. It proves that the system increases task performance speed, improving accuracy in horizontal movements but indifferent in vertical movements compared to unassisted performances.

1 Introduction

The number of automated tasks raises continuously in many industrial working fields, aiming to improve work efficiency and reduce work load. On constructional sites, this is difficult to realize because of considerably changing environmental conditions and customized tasks. However, assistance of craftsmen is possible with respect to repetitive activities, e.g., transport of heavy loads as well as measuring and marking targets.

Augmented reality (AR) is a solution for visual assistance in structured environments. Its virtual information is shown pose correct in real world [Azu97]. By comparing constructional map information with the working setup, a correct display of borehole locations enables successful drilling without damaging wires and pipes underneath plaster. When working without an assistance, craftsmen need much more concentration, time, and effort to fulfill these tasks accurately. Errors can result in pro-

K. Nuelle (✉) · S. Tappe · T. Ortmaier
Institute of Mechatronic Systems, Leibniz University Hanover, Appelstr. 11a, Hanover, Germany
e-mail: nuelle@imes.uni-hannover.de

S. Bringeland · B. Deml
Institute of Human and Industrial Engineering, Karlsruhe Institute of Technology,
Engler-Bunte-Ring 4, 76131 Karlsruhe, Germany

© Springer Nature Switzerland AG 2018
A. Karafillidis and R. Weidner (eds.), *Developing Support Technologies*, Biosystems & Biorobotics 23, https://doi.org/10.1007/978-3-030-01836-8_17

found implications and need to be avoided. AR applications need to track the line of sight in order to display virtual 3D information on head mounted or handheld displays as well as stationary displays or projectors [Kre10]. Many planar projection surfaces exist in construction sites and usually 2D images are used for instructions. Therefore, a line of sight tracking is not necessary, making it a suitable environment for projector base AR. Stationary solutions as in workshops or operation room [Dos16, Wu14] as well as mobile solutions with external tracking system [Gav12, Kob10] are impracticable because of the chancing environment. The presented system solves this problem by self-localization using feature-based simultaneous localization and mapping (SLAM) [Fre10]. Therefore, no extra equipment is needed for assistance.

2 System Design and Accuracy Evaluation

For the outlined necessary support for craftsmen in their daily routine, a visually assistive system is proposed, supporting two tasks. First, it displays borehole positions in correct world locations. Second, it helps the worker to align a drill perpendicular to a wall. In this section, methods used to build and to evaluate the AR system are presented. In addition, the achievable accuracy is obtained.

2.1 System Setup

For displaying additional information to workers, the proposed prototype, see Fig. 1, consists of two main devices: a Microsoft Kinect One® and the laser projector Picopro from Celluon. The Kinect owns a 1080 p rgb camera and a 512×424 infrared depth camera with a maximal framerate of 30 Hz. The projector resolution is 1280×720 p, digitally upscaled to 1920×720 p. ROS is used as software framework handling communication between the components and software modules.

2.2 Visualization of Borehole Positions and Drill Orientation

To visualize borehole positions correctly, the relative orientation and position between the constructional map and the projector is needed. A two-camera calibration [Wie15] is performed between the Kinects rgb and depth camera with a planar calibration pattern to determine the extrinsic, intrinsic and distortional parameters. Additional plane-based projector calibration [Fal08] is realized with respect to the kinects rgb camera. As a result, the extrinsic parameters are stored in transformation matrix $^P T_C \in \mathbb{R}^{4 \times 4}$ and the intrinsic projector parameters in $^{PI} \widetilde{T}_P \in \mathbb{R}^{3 \times 4}$, converting projector coordinates into projector image pixel. A feature-based SLAM approach is used to determine the system's world coordinates. In the first step, the room is scanned

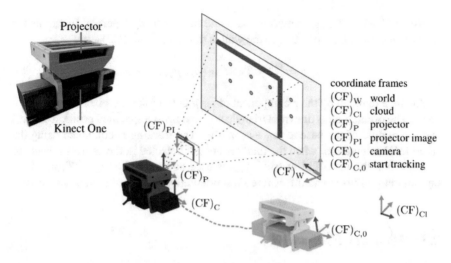

Fig. 1 Prototype of the augmented reality system (left), system setup, and relevant coordinate frames (right)

by the worker running the real-time appearance-based (rtab) SLAM [Lab14] in mapping mode to create a point cloud of the environment. When the system is turned on, it initializes the point cloud coordinate frame $(CF)_{Cl}$ at its current pose. A standardized rule mark on the wall defines the world coordinate frame $(CF)_W$. Its location is known in the constructional map and all information is defined in reference to this frame. To detect it in camera coordinates, template matching on the canny filtered 2D image is used. Point based registration of the corresponding points of the cloud with the map results in the transformation matrix $^{Cl}T_W$ from world to cloud to finish the initialization process.

While working with the system, rtab-SLAM is changed into tracking mode. The transformation matrix $^{C,0}T_{Cl}$ is set when the system localizes itself in the recorded cloud. With respect to $(CF)_{C,0}$, the tracking information is stored in $^{C}T_{C,0}$. Transforming a 3D position $_{(W)}x = (x, y, z, 1)^T$ with homogenous addition in world coordinates into projector image coordinates can be achieved by:

$$_{(PI)}p = {}^{PI}\tilde{T}_P{}^P T_C{}^C T_{C,0}{}^{C,0} T_{Cl}{}^{Cl} T_{W(W)}x = {}^{PI}T_{W(W)}x \text{ with } _{(PI)}p = s \cdot (u, v, 1)^T$$

where s is the scale factor, the introduced transformation matrixes are multiplied. The x and y components of $_{(PI)}p$ are divided by the z component in order to obtain the correct position u and v of the borehole in pixel. It is shown as reticle in the projection image to display boreholes in correct locations.

To specify the drill orientation, a IMU MTi-G-710 GNSS from XSens and an Aruco AR Marker are placed on the drilling machine. The AR coordinate frame is then defined by a printed marker board on the wall. The marker x-axis $_{(AR)}e_x = (p_x, p_y, p_z)^T$ is parallel to the drill and taken from the transformation matrix $\left(^{C}T_{AR}\right)^{-1}{}^{C}T_{Marker}$ between board and marker using OpenCV functions for marker detection. Its projection to the wall is transformed in $(CF)_{PI}$ by:

$$_{(PI)}\boldsymbol{p} = {}^{PI}\tilde{\boldsymbol{T}}_{P}{}^{P}\boldsymbol{T}_{C}{}^{C}\boldsymbol{T}_{AR}\left(p_x, p_y, 0, 0\right)^{T}.$$

Again, a division by the z component of $_{(PI)}\boldsymbol{p}$ determines the pixel coordinates u and v. The orientation is drawn as triangle with the tip at a specified position (u_0, v_0) and the middle of the base at $(u_0 + u, v_0 + v)$. The tracking is continued using the fused IMU orientation, when the markers are not detected in the camera image. To achieve that, the last valid matrix $^{C}T_{AR}$ is used and multiplied by $^{AR}T_{IMU}$ of the current internal fused and offset-free IMU orientation (no translation is assumed).

2.3 Accuracy Evaluation

To evaluate the achievable accuracy of the system, two tests are performed. The first investigates the localization accuracy and the second examines the projection accuracy. A stepwise evaluation procedure is taken to identify error sources. The initial localization within the map is not analyzed because of missing ground truth data.

To evaluate the SLAM tracking accuracy, the system is moved along a known path by a KUKA LBR iiwa 7 R800 with a precision of 0.1 mm, comparing position measurements. While SLAM is running, the robot moves 200 mm along its base axis, stops and returns to its initial pose. This process is repeated for each axis, followed by 90° rotations around each axis. The rotation around the x-axis is only 35° because of singularities. Each robot motion takes 30 s and consists of nine segments, identified by a stop of the robot after each translational and rotational movement. The distance accuracy between the known robot position and the measure SLAM position is calculated after each stop, neglecting the orientation. In all measurements, the tracking is lost due to an insufficient number of features in the camera image when rotating around the horizontal axis and, therefore, capturing the laboratory ceiling. This results in a significant position offset after the system relocalizes. Therefore, rtab-SLAM achieves mean (M) error of 1.0 mm with standard deviation (SD) of 2.3 mm in translational movements but $M = 48.0$ mm, $SD = 4.2$ mm error in rotations, resulting in a mean distance error over 27 measurements of $M = 15.1$ mm, $SD = 22.1$ mm.

A projection error results from inaccuracy in system calibration. To evaluate this error, the system localizes itself via AR marker tracking in the camera image $^{C}T_{AR}$ instead of SLAM, allowing to exclude the previously determined errors. A new world reference is set by a printed AR marker board $(CF)_{AR}$, with four markers in the corners and an empty space in the middle. The system projects another marker board within a square of 10 cm with four corner points $_{(AR)}\boldsymbol{p}_i$ around the center of

$(CF)_{AR}$. For visualization, these points are transformed into projector coordinates by:

$$_{(PI)} p_i = {}^{PI}\tilde{T}_P {}^{P}T_K {}^{K}T_{AR (AR)} p_i .$$

The projected marker board is affine transformed to fit into $_{(AR)} p_i$, ideally forming a perfect square. For marker tracking, an adaptive filtered camera image is needed because white parts of printed board have the same intensity as black parts of projected markers. Calibration errors as well as AR marker tracking errors result in displacement and incorrect orientation of the projected marker board with respect to the reference marker board. The projection error is determined by calculating the reprojection error of the projected marker and transforming it in AR coordinates. In 19 measurements the projection error is $M = 8.0$ mm, $SD = 4.7$ mm and the orientation error accounts for $M = 1.6°$, $SD = 1.0°$.

3 User Evaluation

To evaluate the system's visual presentation of the drilling orientation, 20 volunteers ($M = 30.45$, $SD = 6.09$ years of age) with professional experience in the handling of drilling machines are introduced to the system. They repeat two simulated drilling activities once with and once without visual guidance evaluating the support in aligning a drill. Thereby, the following two were studied:

(a) The visual system appears to have difficulties in perceiving the spatial orientation of two objects correctly. Thereby, especially oblique angles are often not estimated precisely [Orn07]. For this reason, it is expected that the drilling machine with the visual assistance is adjusted straighter and faster into the final position. In addition, the orientation should be more orthogonally.
(b) Further, it is assumed that the mental workload is reduced by the use of the assistance system as, it is no longer afforded to judge depth information or to process the angles of the objects.

In preparation of the study, ten black colored boreholes were painted on a white wall, as the system accuracy of drill hole projection needs to be improved for rotational movements before evaluating it in a user study, see Fig. 2. The boreholes were arranged in two rows with a vertical distance of 10 cm. Each row consists of 5 marks with a horizontal distance of 10 cm to each other. The drilling machine was equipped with a gyroscope (Aukru MPU-6050-Module) and a pressure sensor (SunFounder Button Module) on the activation button of the machine to assess the users' handling behavior. Both sensors were calibrated and synchronized to each other. The gyroscope was used to track the horizontal and vertical angles between the drilling machine and the boreholes on the wall independently of the IMU and AR marker tracking. The pressure sensor triggered the start and the end time of the drilling activity. After having completed both tasks the participants were asked

Fig. 2 Sketched set up of
the user study

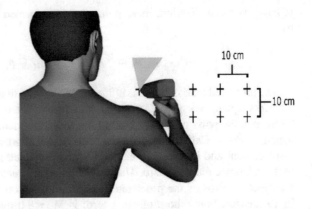

to fill out the NASA Task Load Index (NASA-TLX) questionnaire [Har88] to rate
their mental workload as well as the User Experience Questionnaire (UEQ) [Lau08].
The UEQ assesses both task-related, pragmatic quality aspects (Perspicuity, Efficiency, Dependability) and not task-related, hedonic quality features (Stimulation,
Originality) of products.

In the user study, it turned out that the horizontal angles between the drilling
machine and the boreholes do not differ significantly with regard to the final position
irrespective of whether the AR system has been used or not (T = 58, $p = 0.08$, r =
−0.39). Although it is to be mentioned that, the mean horizontal angle was lower
(M = 1.85°, SD = 0.93°) when working without visual guidance compared to the
visual guidance condition (M = 1.23°, SD = 0.74°). For the mean vertical angle a
significant effect can be reported, whereby significantly lower values (T = 161, $p =
0.04$, r = 0.47) are to be observed with visual guidance (M = 1.36°, SD = 1.37°)
than without guidance (M = 2.3°, SD = 2.24°). When working with the AR-system,
the participants need more time (T = 35, $p = 0.01$, r = −0.58) for each borehole
(M = 5.15 s, SD = 1.43 s) compared to the condition without guidance (M = 4.45 s,
SD = 1.20 s). Furthermore, the subjective workload ratings (NASA-TLX) do not
differ significantly between both conditions (T = 100, $p = 0.84$, r = 0.40). Despite
of this, the augmented reality system is rated positively along all six scales of the
UEQ, see Fig. 3.

Fig. 3 The AR system
compared to the products in
the benchmark data set from
the Excel tool
(ueq-online.org; [Sch14])

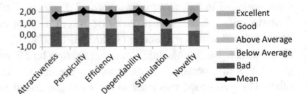

4 Discussion

A mobile AR system is presented for projecting drill holes in correct position on a wall and, additionally, helping craftsmen to align a drill perpendicular to a wall. It operates with internal sensors only and was developed as module for an exoskeleton [Bri16] to enable haptic and visual guidance but can be used as standalone device. The system accuracy, user acceptance and usability of the assistance system is evaluated.

The location accuracy depends mainly on the number of features in the environment. The rtab-SLAM achieves high accuracy in translations, but rotating movements are tracked with less precision. The high rotational error mainly occurs in low feature environment and results from inaccurate relocation after lost tracking. The fusion of feature and depth-based SLAM might solve this problem. The projection rate depends on the frame rate of the Kinect (30 Hz), but it also includes a noticeable delay caused by data processing. Fast camera movements result in disturbing jumping motions of drill hole positions. When adding an IMU to the system, camera pose and imu data can be fused by integrating acceleration data twice and updating with the camera pose in a Kalman filter to smooth jumping movements.

In the user study, the boreholes were located below the shoulder heights, the participants forcing them to look downwards to observe the drilling machine. Due to this perspective of view, the vertical angle may be more difficult to estimate than the horizontal angle. Thus, it may be reasonable that the participants used the visual guidance primarily for the vertical positioning of drilling machine. As the augmentation was displayed above the boreholes several comparing gaze movements between the augmentation and the drilling machine were needed; this may have caused longer positioning times. The additional information from the visual guidance did not cause a higher variability in the movements to the boreholes. According to that, the augmentation system is not leading to any additional distractions influencing the movement of the drilling machine as well as the experiences workload by the participants. Furthermore, the duration of the complete task has to be considered in the analysis of the subjective ratings. The tasks may be too short to cause all possible implications which occur using the system in a real working environment with regular working times. To sum up, the visual guidance improves the vertical positioning accuracy. The location of the augmentation should be close to the borehole in order to reduce the increased gaze times.

Acknowledgements This research received funding by the Federal Ministry of Education and Research of Germany (BMBF 16SV6175).

References

[Azu97]　Azuma, R. (1997). A Survey of Augmented Reality. *Presence: Teleoperators and Virtual Environments, 6*, 355–385.

[Bri16]　Bringeland, S., Heine, T., Hoffmann, M., Stein, T., & Deml, B. (2016). Ergonomische Evaluation eines Handwerker-Kraftassistenzsystems, Arbeit in komplexen Systemen—Digital, vernetzt, human?!. *Tagungsband 62. Frühjahrskongress der Gesellschaft für Arbeitswissenschaft*. GFA Press, Dortmund.

[Dos16]　Doshi, A., Smith, R., Thomas, B. H., & Bouras, C. (2016). Use of projector based augmented reality to improve manual spot-welding precision and accuracy for automotive manufacturing. *Advanced Manufacturing Technology*, 1–15.

[Fal08]　Falcao, G., Hurtos, N., & Massich, J. (2008). Plane-based calibration of a projector-camera system. *VIBOT Master*, 1–12.

[Fre10]　Frese, U., Wagner, R., & Röfer, T. (2010). A SLAM Overview from a User's Perspective. *KI - Künstliche Intelligenz, 24*, 191–198, 2010.

[Gav12]　Gavaghan, K., Anderegg, S., Peterhans, M., Oliveira-Santos, T., & Weber, S. (2012). Augmented reality image overlay projection for image guided open liver ablation of metastatic liver cancer. *Lecture Notes in Computer Science: Augmented Environments for Computer-Assisted Interventions 7264* (pp. 36–46). Berlin, Heidelberg: Springer.

[Har88]　Hart, S. G., & Staveland, L. E. (1988). *Development of NASA-TLX (Task load index). Results of empirical and theoretical research. Human mental workload (Advances in psychology)* (Vol. 52, pp. 139–183). Elsevier.

[Kob10]　Kobler, J.-P., Hussong, A., & Ortmaier, T. (2010). Mini-Projektor basierte Augmented Reality für medizinische Anwendungen. *Tagungsband der 9. Jahrestagung der Deutschen Gesellschaft für Computer- und Roboterassistierte Chirurgie e.V. (CURAC)*, (pp. 115–118).

[Kre10]　Van Krevelen, D., & Poelman, R. (2010). A Survey of Augmented Reality Technologies. Applications and Limitations. *The International Journal of Virtual Reality, 9*, 1–20.

[Lab14]　Labbé, M., & Michaud, F. (2014). Online global loop closure detection for large-scale multi-session graph-based SLAM. In *Proceedings of the IEEE/RSJ International Conference on Intelligent Robots and Systems*, (pp. 2661–2666).

[Lau08]　Laugwitz, B., Held, T., & Schrepp, M. (2008). Construction and evaluation of a user experience questionnaire. In A. Holzinger (Ed.), *Proceedings of USAB'08* (pp. 63–76). Berlin, Heidelberg, Springer.

[Orn07]　Ornkloo, H., & von Hofsten, C. (2007). Fitting objects into holes. On the development of spatial cognition skills. *Developmental psychology, 43*(2), 404–416.

[Sch14]　Schrepp, M., Hinderks, A., & Thomaschewski, J. (2014). Applying the user experience questionnaire (UEQ) in different evaluation scenarios. In A. Marcus (Ed.), *Design, user experience, and usability. Theories, methods, and tools for designing the user experience. DUXU. Lecture Notes in Computer Science 8517*. Cham: Springer.

[Wie15]　Wiedemeyer, T. (2015) IAI Kinect2. https://github.com/code-iai/iai_kinect2. Institute for Artificial Intelligence, University Bremen. Accessed June 12, 2015.

[Wu14]　Wu, J., Wang, M., Liu, K., Hu, M., & Lee, P. (2014) Real-time advanced spinal surgery via visible patient model and augmented reality system. *Computer Methods and Programs in Biomedicine, 113*, 869–881.

Comprehensive Heuristic for Research on Assistance Systems in Organizational Contexts

Daniel Houben, Annika Fohn, Mario Löhrer, Andrea Altepost, Arash Rezaey and Yves-Simon Gloy

Abstract In this paper, we develop heuristics to examine social and organizational factors which influence the success of the introduction of assistive systems in organizational contexts. To this end, we draw on the theory of negotiated order and the approach of zones of uncertainty and apply them to socio-technical arrangements. We present the heuristics on the basis of some qualitative interviews. Finally, we discuss the potential of our heuristics for research on assistive systems in general.

1 Introduction

Whenever new technologies make their way into the workplace, questions regarding the composition of (technical) work and its (re-)structuring and organization in society are raised anew. Digitization—usually understood as the use of interconnected, partly intelligent, assistive or self-controlling tools—is currently discussed as such a new challenge and oftentimes referred to as a "new technology revolution" which follows directly from the previous advances made on the basis of information and

D. Houben · A. Fohn (✉)
Institute for Sociology, RWTH Aachen University, Aachen, Germany
e-mail: afohn@soziologie.rwth-aachen.de

D. Houben
e-mail: dhouben@soziologie.rwth-aachen.de

M. Löhrer · A. Altepost · A. Rezaey · Y.-S. Gloy
Institute for Textile Engineering, RWTH Aachen University, Aachen, Germany
e-mail: mario.loehrer@ita.rwth-aachen.de

A. Altepost
e-mail: andrea.altepost@ita.rwth-aachen.de

A. Rezaey
e-mail: arash.rezaey@ita.rwth-aachen.de

Y.-S. Gloy
e-mail: yves.gloy@ita.rwth-aachen.de

© Springer Nature Switzerland AG 2018
A. Karafillidis and R. Weidner (eds.), *Developing Support Technologies*, Biosystems & Biorobotics 23, https://doi.org/10.1007/978-3-030-01836-8_18

communication technologies [Kuh14, p. 123]. In that respect, Industrie 4.0 seems to be on everyone's lips and addresses the progressive digitization and socio-technical integration of communication, organization, and work in modern industrial production at various levels. From a technical point of view, this involves the increasing integration of all or almost all parts of the production process and the integration of information systems and complex information resources into a consistent digital form of information [Boe96, p. 110]. Industrial digitization must thus be understood as a socio-technical process of the dissemination, production and use of digital technologies in completely different sectors of public and private sector organizations, which is closely connected to economic and social orders, and in particular the change of post-Fordist forms of production [Nac15].

It is not certain, however, whether we are dealing with a significant new leap in technization or with a simple continuation and intensification of existing technology in which information has become the essential core element of production [Boe16, p. 34]. Paradoxically, though, it is not easy to make a realistic assessment of digitization in the workplace. On the one hand, it is undisputed that the exponential increase in the performance of computational capacities has led to far-reaching changes both in private life and work [Hou17]. On the other hand, we are faced with a multitude of exaggerated expectations and politically bolstered narratives like Industrie 4.0. In the course of this process, for instance, new interrelationships between technical and social structuring have been propagated and a substantial reorganization of well-established socio-economic configurations, organizational and network structures are promised almost universally. However, the exact nature of these new interdependencies is usually quite uncertain, therefore, analyses range from stressing comprehensive decentralization and democratization potentials to warnings of strong centralization and power struggles [Kit14].

What do we know about those developments and how can the state of current research be summarized? Many recent studies on the technological advancement of work have focused on questions of standardization, precarization, subjectivization, intensification, flexibilization, and delimitation [Tri11, p.609], which are discussed under the keywords Industrie 4.0 and crowdwork [Hir15]. Mostly, it is assumed that the above-mentioned trends will continue to intensify [Boe14], in particular that the requirements for personal responsibility, self-regulation and self-management will increase significantly due to more complex interfaces, manifold feedback processes, increasing interaction and acceleration of production processes [Kuh14, p.123f.]. Concerning the subject of work, there was hope that industrial automation and digitization would lead to a "post-capitalist" society [Mas16, Srn15]. In terms of organization, new technologies were often regarded as an essential prerequisite for flat hierarchies [Sat15]. On the other hand, the omnipresence of digital sensor technologies in production was interpreted as a revitalization of classical Taylorism and a radicalization of surveillance [Zub15, Mey17].

As we suggest, a comprehensive approach to the study of digitization and particularly to the implementation of assistive systems within organizational contexts must take into account the process-like nature of technological development and the complexity of application contexts instead of focusing on the mere functional promises of an alleged "next technological revolution".

Organizational research has shown that formal processes of change, such as a top-down implementation of digitization, always correspond to an informal substructure in the company social order. Both are in permanent development and related to each other, although initially only the formal part is directly visible [Mey77]. Hence, the introduction of assistance systems to industrial production processes is also likely to stir up socio-technical structures on a firm level in a way that so far only could be researched insufficiently. Additionally, as brought forward by Science and Technology Studies (STS), social and sometimes even political programmatics are inscribed into technology and socio-technical routines [Bau17]. However, there is continuing controversy over how these political qualities actually develop in specific organization contexts. In this respect, we stress the need to investigate the micro-political contingency of digital assistance on the firm level. Hence, in what follows we develop an analytical heuristic to tend to these social and organizational aspects.

2 Context: Challenges of the German Textile Industry

Our theoretical approach is grounded on the notion that social and micro-political mechanisms are no less important to the implementation of assistance systems than the objective technological aspects—regardless of whether the introduction of assistance systems involves strong or incremental alterations to the industrial processes in question. To discuss social factors in the introduction of production assistance systems, in this essay, we object to the ontological assumption that social order is inherently stable. Instead, socio-technical systems are considered a dynamic outcome of continuous social interactions that consolidate or challenge given socially negotiated orders. To highlight that fact and to develop a corresponding heuristic, we employ Strauss's concept of negotiated order and the micro-political perspective of zones of uncertainty elaborated by Crozier and Friedberg.

2.1 Negotiated Socio-Technical Order

The theory of negotiated order rejects that social orders—and by extension industrial companies or organizations in general—are to be seen as largely passive and thus stable contexts for technological developments and therefore only need to be considered as analytically secondary at best. Instead, we follow Strauss's [Str78] and Fine's [Fin84] suggestions that what we commonly conceive as order and stability are in fact highly contingent social outcomes themselves. Corresponding to its roots in symbolic interactionism, the central premise of the theory of negotiated order is that institutions and social order must constantly be created, maintained, and reproduced by social interaction. From this point of view, social structures are created through exchanges and negotiations between actors who in turn create and inhabit a social context [Fin84]. The theory of negotiated order thus differs distinctly from theoretical perspectives based on the stability or inertia of social systems. It considers disturbances in any given social or organizational context, such as a change of personnel, progress in technology, or alterations in work routines as triggers for renegotiation or re-evaluation of that very social or organizational context. In extreme cases those changes can be disruptive and might even lead to the creation of a new social or organizational order [Str63, p.167].

To harness the potential of these universal insights for the analysis of technological change, such as the introduction of an assistance system into existing work routines, we propose to expand the meaning of social order to socio-technical order in which technological development is co-determined by social interaction and technical factors that jointly build the social, i.e., organizational structure [Lab12, Dok12].

Due to its contextual focus and its emphasis on social interaction the theory of negotiated order is well suited for the study of social dynamics in firms. In this respect, firms are complex socio-technical contexts that are ultimately influenced by employees as well as their institutional and organizational environment such as other companies, influential external stakeholders, regulatory authorities and, of course, technological trends [Sco08]. A theory that analytically approaches both stability and change in organizations should therefore be able to take different types of actors into account and specify what contextual characteristics influence the relevant social interactions in the respective organization. Hence, our approach suggests a model of social or organizational order that is maintained by a recursive relationship between the given structural context, the specific context of the concrete negotiations and other relevant social interactions, and the results of that interaction which feed back into the situational organizational context. Put differently, by addressing those aspects the theory of negotiated order provides a conceptual framework that allows us to identify the specific conditions for socio-technical change in communities of practice [Wen00]. The challenge for researchers who analyze a specific context, such as socio-technical communities within organizations, is to find out which specific conditions in this structural context influence social interactions and how to explain those. To begin with, the structural context can be identified as the broader institutional conditions under which negotiations take place [Str78, p.98]. This normally includes the respective industry, the legal and regulatory environment, important ideological debates, the predominant culture and authority relationships in the respective industry or field of expertise. Consequently, the structural context is always the background of the specific situational negotiating context. The latter is therefore a subset of the observable characteristics of the broader structural context that take an effect on social interactions and their results [Dok12].

Strauss defines contextual conditions as those contextual properties which provide a legitimate frame of reference in a given social interaction: First, contextual conditions define the types of actors who interact within a negotiating context and the relationships between them. Consequently, second, contextual conditions have a causal impact on social interactions and their outcomes. Third, contextual conditions can impart the causal effect of another contextual condition on the interaction results [Str78]. Hence, contextual conditions that have causal or mediating effects influence social interactions as they suggest or conclude agreements, form coalitions, identify and formulate issues, examine possible solutions, and consequently result in interactions that either reinforce or challenge the existing socio-technical order. Moreover, interaction results are fed back into the structural context and affect the negotiation context [Dok12, Str63, p. 165]. For example, ongoing interactions between individuals lead to a constant development and change of the network of relationships between individuals, regardless of whether there is a radical change in the larger socio-technical order or not. Thus, the results of the interactions are also incorporated into the larger organizational structure. Whereas processes that reinforce the existing socio-technical order normally fosters stability, e.g., incremental changes that are on the established technological path, processes that challenge the socio-technical order can disrupt existing organizational structures and induce technological development [Bec11, Dok12]. In summary, the theory of negotiated order provides a general framework that researchers can use to examine the overall organizational conditions and social processes that challenge or strengthen a given socio-technical order and by that are relevant to the implementation success of new technologies.

2.2 Zones of Uncertainty

Whereas the theory of negotiated order concentrates on the social, organizational, and by extension technological contexts it lacks a comprehensive guideline to the concrete ways and forms of negotiations and their underlying formal and, even more intriguing, informal organizational power structures. Crozier and Friedberg addressed informal power processes within formal corporate structures with the idea of the zones of uncertainty [Cro79].

This concept is based on the initially paradoxical realization that organizational structures and rules create uncertainty and interdependencies. Under a closer look, social power exhibits a certain inherent dialectic since the power of person A stems from the degree of reliance person B feels towards person A. Organizational control is hence a function of the degree of ignorance, uncertainty and interdependency in a situational context. To put it simply: The more the work output of person A, say a blue-collar mechanic, is dependent on the work and devotion of person B, say a system administrator, and simultaneously cannot be controlled by Person A, the more power does person B have over person A. Hence, within any collaborative technological innovation resides the potential to shift the operational balance within any given work context. By inducing new of the forms, i.e., zones of uncertainty, new technologies potentially shake the distribution of power resources within a firm and eventually also between different companies [Fri92]. As a theoretical concept, zones of uncertainty thus offer a tool to account for the dynamic power relations and the negotiations they induce. Therefore, in order to regulate and control their workforce, firms need to manipulate the zones of uncertainty relevant to their production processes. Crozier and Friedberg point out that the principles of order of organizational substructures include those power resources that stem from the mastery of particular zones of uncertainty—for example, from the ability to program digital processes in a given value-added chain as a result of qualifications or inclinations. To this end, Crozier and Friedberg identify four fundamental sources of power in organizations and four distinct types of uncertainty zones: firstly, expertise, mastery of specific knowledge and functional specialization; secondly, relations between organizations and their environment; thirdly, information and communication flows; and fourthly, general organizational rules [Cro79].

We assume that the diffusion of principles of Industrie 4.0 will inevitably give rise to previously largely unknown standards and new working contexts. Hence, if the introduction of new technologies has the potential to alter and re-organize the existing socio-technical order of work in a given industry, it necessarily creates zone of uncertainty which will rattle well-established conventions and work routines, as well as existing employee networks and other work-related demands [Fre13]. Even previous forms of internal cooperation as well as cross-company cooperation would be subject to profound changes, so that the devaluation of stable processes, structures, and resources can be examined as a growing number of zones of uncertainty. Consequently, it must be assumed that digitization and the introduction of assistive systems will not simply reproduce but interfere with former industrial organizational structures [Hou18]. Moreover, with Crozier and Friedberg [Cro79], socio-technical zones of uncertainty are induced by digitalization or the implementation of assistive systems would affect many levels of organizational life, namely in transformations of organizational culture, work processes and power relations. In this respect, one important facet of the digitization of work that has only been scarcely researched so far can be found in the changing relationship between workplace formality and informality [Gro15]. Here, new informal [technical] knowledge coupled with the practical forms and resources of social power of a given firm's employees might interfere with the previously legitimate operational processes and dominant social structures.

2.3 Interim Conclusion: The Need for a Participatory Development of Technology

Against this theoretical backdrop that highlights the contingency of socio-technical orders and offers perspectives to identify hidden power relations, we propose that the effectiveness of an assistance system in organizational contexts not only depends on aspects of concrete technological advancement but also on the question of how the firms and their staff, work cooperatively with a particular technology and on how that technology affects their relationship dynamics. Analytically, it is therefore of vital importance to overcome a technological determinism which claims that innovation is determined solely by the characteristics of the technology itself. Instead, it should have become evident that the path of technological development is to a high degree determined socially by the interactions between the workers that form a socio-technological community [Häu14, Wen00]. This premise of an ontological dynamic of socio-technical arrangements, which has already been explained above, has several implications for technology research and the participatory development of new technologies: In the process of challenging and negotiating details of a given assistive system that will eventually impose changes to organizational processes, staff members, stakeholders and other relevant experts might take on the opportunity to challenge the predominant socio-technical order and these challenges could trigger a wider transformation on the organization or even industry level. Thus, social interactions within a technological community have the potential to lead to either more sustainable or radical changes in a given socio-technical order.

3 Understanding the Organizational Base for Assistant Technology in the Textile Industry—Preliminary Results

To set off this idea of participatory development of assistive systems in the German textile industry, an interdisciplinary team of researchers, consisting of specialists from engineering, sociology and education sciences in order to conduct the project "SozioTex—New socio-technical systems in the textile industry", granted by the Federal Ministry of Education and research. The project aims at the development and implementation of a digital assistance system for weaving mills, which supports workers heterogeneous in age, gender, qualification and cultural background in adapting to varying contexts and scenarios concerning Industrie 4.0. By accelerating and simplifying learning and performance, for example, conducting a warp beam change, the research group expects a growth of productivity of around 30%. In order to provide the transfer to further applications and industrial sectors, a main result of "SozioTex" will be the design of a generic procedure model for assistive systems for weaving mills which might serve as a model to enhance the efficiency of industrial production processes in general. The prospective users, managers and employees of three weaving mills in Germany, participated in requirement analysis and design from the very beginning of the project, for example, by taking part in workshops, surveys and group discussions. This participative approach in systems design permits to make use of the expertise of the employees and to gain a better solution. Furthermore, without involving the future users in the design process, the acceptance of technical systems and corresponding organizational arrangements is at risk [Alt17].

In February 2017, a qualitative study in the form of guideline-supported interviews was carried out as part of SozioTex. The organizational conditions of small and medium-sized enterprises in the German textile industry were investigated based on three reference weaving mills in North Rhine-Westphalia. Human resources managers and employees at the higher management level were interviewed about the organizational structures, the degree of automization and the digitization and data fostering tendencies in operational practice. The preliminary findings are presented in the following paragraphs. Yet, to gain a more comprehensive view of the entire organization and to be able to make specific statements, the shop floor levels should also be the object of investigation in the future.

3.1 On the Way to Industrie 4.0—Expectations, Uncertainties, and Limitations

Since the wider development of the industry sets important guidelines, the perception of their respective field by the firms allows for an idea of the wider organizational context. In our interviews, the respondents did not yet see themselves fully committed to Industrie 4.0, but they believed that from a development towards Industrie 4.0 the more problematic production processes that for example pose health risks to employees as well as routine tasks could be automated in the future. They also articulated the hope that employees might have more time for creative tasks. In addition, it was stated that the expansion to Industrie 4.0 could lead to better quality assurance and create benefit for the customer. However, when asked what exactly Industrie 4.0, or rather a weaving mill 4.0 might be, the interviewees could not verbalize it. Nevertheless, the expansion of cooperation with machine manufacturers was classified as a necessity, since opening up the scope of possibilities was regarded as the only way Industrie 4.0 could be implemented in reality. In addition, with regard to the recruitment of junior staff, job attractiveness could be increased by digitization and automization leading to easier work and technological gimmicks that address a new young target group and make the weaving industry attractive again [W2_MA_01].

The prevailing uncertainty towards the upcoming technological—and by that organizational—changes is represented by the statement that the small and medium-sized enterprises we interviewed felt that they have still enough to do with Industrie 3.0. Therefore, they are currently concentrating their automation efforts on advancing paperless work and on improving interface designs. The transfer and storage of data and, in particular, of process information, have the highest priority on their agenda, though. Respondents especially saw potential in automating the machine allocation process, for example, or in digitizing orders, production instructions and routines. The visualization of work processes and the use of RFID tags and barcodes are currently the highest level of development of weaving technology 4.0. In addition, the connection to the Internet of Things and the move towards highly digitized weaving mills is still in its infancy.

3.2 Communication Tools and the Ambivalence of Digital Assistance

Communication and interaction are at the heart of any socio-technical order. There-fore, the way in which the use of assistance systems might influence the communica-tion patterns at work is crucial. In the weaving industry, traditional communication is particularly difficult due to the enormous strains caused by high decibel levels. Since employees need to protect their hearing they are particularly restricted in their intra-organizational acoustic communication abilities. Precisely for this reason, there is great potential for the development of digital assistance tools, which improve the production process, communication and general interaction between employees. Yet, the digitization of internal company communications creates both advantages and disadvantages for employees and thus may have an ambivalent effect, for it closes zones of uncertainty but at the same time may create a panopticon, i.e., a system of permanent probation [Boe08]. Digital technologies—more than analogue technolo-gies—offer opportunities of intervention and require an ergonomic design as well as an exceptionally active adoption by the user [Pau05], which takes place in an inter-play of formal structures and processes with informal practice on the one hand and on the basis of institutionally assigned or informally accumulated power resources on the other hand [Hou17]. Even highly qualified employees who are exposed to new technologies at work are also confronted with new experiences of uncertainty and oftentimes fear to become replaceable [Boe16]. Since industrial processes are highly dependent on power relations and the specific organizational practices of employees, formal and informal processes in the digitized company interact with each other in a complex fashion [Fun08].

3.3 Practical, Implicit, and Informal Knowledge

Specific features of the machinery or certain parameters of the production process as well as information on the respective skills of the workforce are mostly memorized in form of practical, implicit and informal knowledge by employees. However, these forms of knowledge are of the utmost relevance for entrepreneurial success. An aging workforce and the demission of knowledge into retirement will cause severe prob-lems not only for the companies but for the whole industry as well. The possibilities for organizational innovations and space for maneuver in digitization are limited by the necessity of permanent presence of employees. The machines set the pace in the weaving mills and the workplace can only be left out of sight to a very limited extent. In this case, it is necessary to look for automation possibilities together with the employees in a participatory approach. Consequently, the storage, processing, and use of organizational data and its use in corresponding contexts will become one of the central tasks for the future. Here, like in the development of assistance in com-munication (Sect. 3.2), there are corresponding zones of uncertainty and imbalances in power relations because the management is not aware what kind of knowledge should be institutionalized and transferred in order to increase the firm's success [Alt17].

3.4 Negotiation Configurations

Further uncertainties can be found concerning the questions how to implement new technologies and which actors should participate in the process. For instance, the role of the trade unions still seems to be an unresolved issue, since the possibilities of digitization and automization in the working context are only partially accessible and at the same time conceivable for the interviewees. What the respondents are already aware of at this stage, however, is that within the scope of changes leading up to Industrie 4.0, data volumes are being produced which terms of use will have to be clarified in the future. Particularly in this context, it is essential to involve established work councils at an early stage, even if personal data are not recorded. In the interviews, it is criticized that "safety requirements for machines [...] are increasing even more" and "laws are being tightened even more tightly". All in all, structural deficits are identified, which hinder a smooth implementation. Therefore, hope is placed among politics to create more a more suitable legal framework in the future [W2_MA_01]. In addition to that, the interviewees were of the opinion that when it comes to implementing technical innovations "the initiative must be taken by the superiors" [W2_MA_01]. However, implementation processes of technical systems that are initiated top-down could also lead, for example, to employees refusing to accept a top-down assistance system. This would result in a changeover in the balance of power on an organizational level, so that the employees could reject imposed support out of protest and thus exercise power over the management level. They would argue that the management level was anxious to increase efficiency and reduce costs rather than to improve the work situation for their employees.

4   Conclusion

Our preliminary results show that the reference to the concepts of negotiated order and zones of uncertainty provide insights which should prove useful to the development and design of socio-technical assistance systems. Based on the premise that the employees have both informal problem-solving skills in dealing with the implicit zones of uncertainty induced by the introduction of assistance systems and are able to identify important issues regarding current and future challenges, their informal knowledge and skills can be reconstructed empirically in the further course of our project. This knowledge then must be employed to stress their ability to connect with established manufacturing work and the knowledge of organizational problems built up within it. However, we are well aware that we have only presented a first layout which will have to be worked out in the future.

In this way, the research interest is directed both at informal knowledge about the digital manufacturing process based on the actors' respective operational practice and at the possibilities of a reciprocal knowledge transfer between the relevant actor groups. This approach should enable us to gain a more complete view on potentials and risks of assistance systems in the textile industry, i.e., on design options and the need for regulation. Informal knowledge is mainly about the knowledge of operational processes which is anchored in the hands-on experience of the personnel and which can be used to identify actual options or risks of deployment of digital assistance systems. In particular, the knowledge of the techniques used in everyday production is to a high degree practical, implicit and informal, therefore it may not necessarily be cognitively retrievable and by that it is hard to pass on to colleagues or to transform into digitized programs.

Further, it can be assumed that the issues we discussed for the design of working arrangements will remain relevant for a successful implementation of any assistive system in any organizational context. This calls for empirical research on the practical implementation of assistance systems, their technical form, and their organizational contexts as well as individual coping especially with regard to the related zones of uncertainty for both employees and employers.

Finally, we want to use this insight to encourage all stakeholders to reflect on the contextual conditions and potential challenges to existing socio-technical orders of assistance systems not only in the textile industry. The framework presented here is also intended as a contribution to a new transdisciplinary research paradigm that helps technological progress to meet social, organizational, economic and technological challenges. In this respect, we hope that the heuristics and preliminary results outlined in this paper will contribute to an interdisciplinary dialogue and play their part in creating assistance systems that people really want and need [Wei15, Kar17].

Acknowledgements We thank the German Federal Ministry of Education and Research for the funding of our research.

References

[Alt17]   Altepost, A., Löhrer, M., Ziesen, N., Saggiomo, M., Strüver, N., Houben, D., et al. (2017). Sociotechnical systems in the textile industry: Assistance systems for industrial textile work environment. *i-com, 16*(2), 153–164.

[Bau17]   Bauer, S., Heinemann, T., & Lemke, T. (Eds.). (2017). *Science and technology studies—Klassische Positionen und aktuelle Perspektiven*. Berlin: Suhrkamp.

[Bec11]   Bechky, B. A. (2011). Making organizational theory work: Institutions, occupations, and negotiated orders. *Organization Science, 22,* 1157–1167.

[Boe08]   Boes, A., & Bultenmeier, A. (2008) Informatisierung – Unsicherheit – Kontrolle. Analysen zum neuen Kontrollmodus in historischer Perspektive. In K. Dröge & W. Menz (Eds.), *Rückkehr der Leistungsfrage. Leistung in Arbeit, Unternehmen und Gesellschaft* (pp. 59–90). Berlin: Edition Sigma.

[Boe14]   Boes, A., Kämpf, T., Langes, B., & Lühr, T. (2014). Informatisierung und neue
            Entwicklungstendenzen von Arbeit. *Arbeits- und Industriesoziologische Studien*,
            7, 5–23.
[Boe16]   Boes, A., Kämpf, T., Bultemeier, A., & Lühr, T. (2016). Die Digitalisierung braucht
            den Menschen. In Daimler und Benz Stiftung (Ed.), *Digitale Arbeitswelt – Folgen
            für Arbeit und Gesellschaft, Dokumentation des 14* (pp. 4–13). Innovationsforums
            der Daimler und Benz Stiftung.
[Boe96]   Boes, A. (1996). Subjektbedarf und Formierungszwang. Überlegungen zum
            Emanzipationspotential der Arbeit in der „Informationsgesellschaft". In E. Bul-
            mahn (Ed.), *Informationsgesellschaft - Medien – Demokratie* (pp. 109–124). Mar-
            burg: BdWi-Verlag.
[Cro79]   Crozier, M., & Friedberg, E. (1979). *Macht und Organisation: Die Zwänge kollek-
            tiven Handelns*. Königstein: Athenäum.
[Dok12]   Dokko, G., Nigam, A., & Rosenkopf, L. (2012). Keeping steady as she goes: A
            negotiated order perspective on technological evolution. *Organization Studies, 33*,
            681–703.
[Fin84]   Fine, G. A. (1984). Negotiated orders and organizational cultures. *Annual Review
            of Sociology, 10*, 239–262.
[Fre13]   Frey, C., & Osborne, M. A. (2013). *The future of employment: How susceptible
            are jobs to computerization?*. Oxford: Oxford University Press.
[Fri92]   Friedberg, E. (1992). Zur Politologie von Organisationen. In W. Küpper & G.
            Ortmann (Eds.), *Mikropolitik* (pp. 39–53). Opladen: Westdeutscher Verlag.
[Fun08]   Funken, C., & Schulz-Schaeffer, I. (Eds.). (2008). *Digitalisierung der Arbeitswelt.
            Zur Neuordnung formaler und informaler Prozesse in Unternehmen*. Wiesbaden:
            Springer.
[Gro15]   von Groddeck, V., & Wilz, M. (Eds.). (2015). *Formalität und Informalität in Organ-
            isationen*. Wiesbaden: Springer VS.
[Häu14]   Häußling, R. (2014). *Techniksoziologie*. Konstanz: UTB.
[Hir15]   Hirsch-Kreinsen, H., Ittermann, P., & Niehaus J. (Eds.). (2015). *Digitalisierung
            industrieller Arbeit. Die Vision Industrie 4.0 und ihre sozialen Herausforderungen*.
            Baden-Baden: Nomos.
[Hou17]   Houben, D. (2017). Von Ko-Präsenz zu Ko-Referenz – Das Erbe Erving Goff-
            mans im Zeitalter digitalisierter Interaktion. In M. Klemm & R. Staples (Eds.),
            Leib und Netz (pp. 15–31). Wiesbaden: Springer VS.
[Hou18]   Houben, D., & Prietl, B. (Eds.). (2018). *Datengesellschaft. Einsichten in die
            Datafizierung des Sozialen*. Bielefeld: Transcript.
[Kar17]   Karafillidis, A. (2017). Synchronisierung, Kopplung und Kontrolle in Netzwerken.
            Zur sozialen Form von Unterstützung und Assistenz. In P. Biniok & E. Lettkemann
            (Eds.), *Assistive Gesellschaft, Öffentliche Wissenschaft und gesellschaftlicher Wan-
            del* (pp. 27–58). Wiesbaden: Springer VS.
[Kit14]   Kitchin, R. (2014). *The data revolution*. London: Sage.
[Kuh14]   Kuhlmann, M., & Schumann, M. (2014). Digitalisierung fordert Demokratisierung
            der Arbeitswelt heraus. In R. Hoffmann & C. Bogedan (Eds.), *Arbeit der Zukunft:
            Möglichkeiten nutzen - Grenzen setzen* (pp. 122–140). Frankfurt/M.: Campus.
[Lab12]   Labatut, J., Aggeri, F., & Girard, N. (2012). Discipline and change: How technolo-
            gies and organizational routines interact in new practice creation. *Organization
            Studies, 33*, 39–69.
[Mas16]   Mason, P. (2016). *Postcapitalism: A guide to our future*. London: Macmillan.
[Mey17]   Meyer, U., Schaupp, S., & Seibt, D. (2017). *Digitalized industries: Between dom-
            ination and emancipation*. Call for Papers.
[Mey77]   Meyer, J. W., & Rowan, B. (1977). Institutionalized organizations: Formal structure
            as myth and ceremony. *American Journal of Sociology, 83*, 340–363.
[Nac15]   Nachtwey, O., & Staab, P. (2015). Die Avantgarde des digitalen Kapitalismus.
            Mittelweg, 6, 59–84.

[Pau05]   Paulitz, T. (2005). *Netzsubjektivität/en. Konstruktionen von Vernetzung als Tech-nologien des sozialen Selbst: Eine empirische Untersuchung in Modellprojekten der Informatik*. Münster: Westfälisches Dampfboot.

[Sat15]   Sattelberger, T., Welpe, I., & Boes, A. (Eds.). (2015). *Das demokratis-che Unternehmen: Neue Arbeits-und Führungskulturen im Zeitalter digitaler Wirtschaft*. Haufe: Freiburg.

[Sco08]   Scott, R. W. (2008). *Institutions and organizations. Ideas and interests*. Thousand Oaks: Sage.

[Srn15]   Srnicek, N., & Williams, A. (2015). *Inventing the future. Postcapitalism and a world without work*. London: Verso.

[Str63]   Strauss, A., Schatzman, L., Bucher, R., Erhrlich, D., & Sabshin, M. (1963). The hospital and its negotiated order. In E. Friedson (Ed.), *The hospital and modern society* (pp. 147–169). New York: Free Press.

[Str78]   Strauss, A. (1978). *Negotiations: Varieties, contexts, processes and social order*. San Francisco: Jossey-Bass.

[Tri11]   Trinczek, R. (2011). Überlegungen zum Wandel von Arbeit. *WSI-Mitteilungen, 1*, 606–614.

[W2_MA_01]   Unpublished interview transcript: Werk 2, Mitarbeiter 01. Institut für Soziologie. RWTH Aachen University.

[Wei15]   Weidner, R., Redlich, T., & Wulfsberg, J. (Eds.). (2015). *Technische Unter-stützungssysteme*. Berlin: Springer.

[Wen00]   Wenger, E. (2000). *Communities of practice. Learning, meaning, and identity*. Cambridge: Cambridge University Press.

[Zub15]   Zuboff, S. (2015). Big other: Surveillance capitalism and the prospects of an infor-mation civilization. *Journal of Information Technology, 30*, 75–89.

Human Motion Capturing and Activity Recognition Using Wearable Sensor Networks

Gabriele Bleser, Bertram Taetz and Paul Lukowicz

Abstract Wearable sensor networks enable human motion capture and activity recognition in-field. This technology found widespread use in many areas, where location independent information gathering is useful, e.g., in healthcare and sports, workflow analysis, human-computer-interaction, robotics, and entertainment. Two major approaches for deriving information from wearable sensor networks are in focus here: the model-based estimation of 3D joint kinematics based on networks of inertial measurement units (IMUs) and the activity recognition based on multimodal body sensor networks using machine learning algorithms. The characteristics, working principles, challenges, potentials, and target applications of these two approaches are described individually and in synergy.

1 Introduction

In the nineties cheap motion sensors together with low power, compact wireless processing and communication capabilities started becoming available. This led to the idea of using such sensors for in-field (also called "in the wild") capture of human motion in terms of 3D kinematics [Pic17] and recognition of general human activity [Sch99, Bao04]. Driven by continuous cost, size, power consumption reduction, and integration into accessories and smart textiles [Zhe14], this technology found

G. Bleser (✉) · B. Taetz
Department of Computer Science, Technische Universität Kaiserslautern, Gottlieb-Daimler-Str. 48,
67663 Kaiserslautern, Germany
e-mail: bleser@cs.uni-kl.de

B. Taetz
e-mail: taetz@cs.uni-kl.de

P. Lukowicz
German Research Center for Artificial Intelligence, Trippstadter Str. 122, 67663 Kaiserslautern, Germany
e-mail: Paul.Lukowicz@dfki.de

© Springer Nature Switzerland AG 2018
A. Karafillidis and R. Weidner (eds.), *Developing Support Technologies*, Biosystems & Biorobotics 23, https://doi.org/10.1007/978-3-030-01836-8_19

widespread use in many areas, where "in vivo" information gathering is important. This ranges from healthcare and sports [Shu14, Won15, Che16, Men16, Ios16] over industrial ergonomics and workflow analysis [Vig13, Won15, Ble15a] to human-computer-interaction and robotics, e.g., [Tag14], to name some prominent examples.

On an abstract level two general approaches can be identified for deriving different types of information from wearable sensor networks [Lop16]. The first approach focuses on the estimation of 3D joint kinematics, which in essence amounts to the capture of the poses (orientations, positions), (angular) velocities, and (angular) accelerations of each relevant body part (body segments or joints). Here, the goal is to enable personalized biomechanical analyzes outside the lab and at relatively low cost, but with comparable accuracy to laboratory-based gold standard systems (e.g., [Sut02]). This approach generally relies on inertial measurement units (IMUs) in combination with model-based sensor fusion algorithms, e.g., [Mie16]. It uses physical and biomechanical models and is independent of training data. State of the art methods typically require one IMU on each body segment that should be captured. In other words, the price for an exact motion estimation (which is suitable for biomechanical analyzes) is the need for a potentially large number of sensors, and possibly strict placement and attachment constraints may need to be observed.

The second approach abstracts from the capture of exact body motions and focuses on using machine learning techniques to build statistical models of relevant activities based on signals from fewer sensors, in particular sensors placed on fewer (often just one) body locations. Here, lesser accuracy and level of detail, dependence on training data, and a "black box" statistical character of the model are the price that has to be paid for a less obtrusive, easier to deploy system. While IMUs also play an important role in this approach, they are often complemented by other sensors ranging from microphones over textile stretch sensors, capacitive body sensors, pressure sensors and ultrasonic sensors to eye trackers and wearable cameras [Shu14, Won15, Ble15a, Pap17].

The following sections describe the individual working principles, challenges, and potential applications of these two approaches, then their existing and potential synergies on method and application level. Finally, different aspects of how the technology can be beneficially used in the context of support systems are summarized.

2 IMU Based 3D Kinematics Estimation

2.1 Working Principles

In the area of IMU based kinematics estimation the motions of a person are approximated through the motion of a pre-known biomechanical model that is driven by noisy and biased IMU measurements (angular velocities, accelerations, mostly also magnetic fields) through a stochastic sensor fusion algorithm. This is in contrast to optical gold standard systems, e.g., [Sut02], where the 3D positions of reflective markers precisely placed on anatomical landmarks are measured directly and joint centers and angles are geometrically derived from these [Lea07].

The biomechanical model typically consists of rigid segments, approximating the human bones. These are connected through joints that can optionally be constrained regarding their degrees of freedom (DoF). Besides a personalized biomechanical model (e.g., in terms of segment lengths), the reconstruction of biomechanically valuable joint kinematics data requires knowledge about the relative transformations between IMU and segment coordinate systems, the so-called IMU-to-segment calibrations. Figure 1 illustrates the above-mentioned aspects.

A sensor fusion algorithm here denotes a combination of a set of stochastic equations to describe the estimation problem, often called a state-space model, and an estimation method to solve this problem. The state-space model defines (1) the vari-

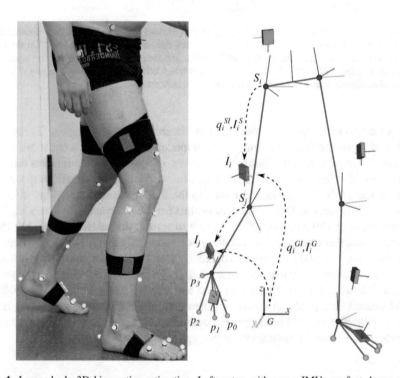

Fig. 1 Lower body 3D kinematics estimation. Left: setup with seven IMUs on feet, lower and upper legs, and pelvis as well as reflective markers according to Leardini et al. [Lea07]. Right: Biomechanical model of the lower body with connected segments (magenta lines), joint centers (red spheres), four contact points on each foot (green spheres), IMU placement, and involved coordinate frames. A technical coordinate system is associated to each IMU (I). The segment coordinate systems (S) are drawn at the proximal ends of the segments. The six degrees of freedom (DoF) transformations, each in terms of an orientation (quaternion) q^{SI} and a translation I^S, between the IMU coordinate frames and the associated segment coordinate frames are called IMU-to-segment calibrations. One such calibration is shown at the right thigh. The symbol G denotes the global coordinate system. The figures have been taken from [Ble17, Mie17]

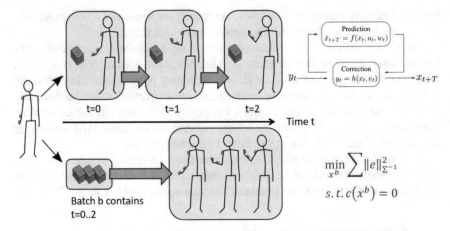

Fig. 2 Illustration of an extended Kalman filter (upper row) and a sliding-window optimization (lower row) based solution to IMU based kinematics estimation of an upper limb. On the right side, a general state-space formulation with motion and measurement models for the prediction and correction step of a recursive filter, as well as, a weighted least squares cost function with hard constraints for batchwize numerical optimization are indicated

ables (states) of interest, i.e., the segment kinematics or joint angles, (2) the evolution of these variables over time (motion models), i.e. difference equations based on assumptions on how the human body moves, (3) how the measurements relate to these variables (measurement models), i.e., forward kinematic equations that relate the motion of the biomechanical model to the IMU measurements. This information is often combined with further constraints from the biomechanical model, such as limited joint DoFs and ranges of motion to restrict the solution space. For IMU based kinematics estimation the resulting estimation problem is nonlinear. Methods to solve this problem (based on noisy data and uncertain assumptions) typically utilize Bayesian inference, where a nonlinear maximum a posteriori estimate can be found in multiple ways [Thr05, Gus12], e.g., via an extended Kalman filter (EKF), which works based on a predictor-corrector scheme, or via sliding-window/moving horizon (nonlinear weighted least squares) optimization, to name two (online-capable) approaches. Figure 2 illustrates these two approaches.

2.2 Challenges and Solution Approaches

IMU based pose estimation typically suffers from integration drift. This is caused by integrating the noisy and biased gyroscope measurements to obtain orientation changes and by using this for gravity-compensating the accelerometer measurements, which are then double integrated to obtain position changes [Kok16]. Orientation drift is often compensated for by additionally using magnetometers. These provide valuable orientation information in the case of a homogeneous magnetic field. However, the assumption concerning the global magnetic field is often violated, particularly in indoor environments [Lig16]. Therefore, recent research addressed the development of new sensor fusion methods, which can work without using magnetometer information; e.g., Miezal et al. [Mie16] showed that a combination of a redundant biomechanical model definition with biomechanical and kinematic constraints accounted for in an optimization-based state estimation method show lower orientation drift and higher biomechanical model error tolerance compared to the more classical kinematic chain and EKF based sensor fusion method. It was also shown that pairwise kinematic constraints can reduce drift at the joints, e.g., [Wen15, Fas17]. To obtain long-term stable global heading orientation estimation (i.e., transversal plane rotation), additional sensors (e.g., cameras [Ble09]) or scenario-dependent assumptions (e.g., a reset pose or walking on a straight line) can be exploited.

Translation drift can be reduced through so-called zero velocity updates at stationary points on the biomechanical model, a well-known concept from the field of Pedestrian Dead-Reckoning [Har13]; e.g., in [Mie17] a probabilistic kinematics-based ground contact estimation method using four contact points on an anatomically motivated foot model (see Fig. 1) was proposed and in [Ble17] it was integrated with different sensor fusion methods. The results show significant drift reduction for different types of locomotion, such as walking, running, and jumping. To obtain long-term stable global translation estimation, again, additional sensors, such as cameras [Ble09], ultrawideband or global positioning system (GPS) [Hol11, Kok15], can be used. Another area of research addresses methods for obtaining valid and reliable IMU-to-segment calibration parameters (see Fig. 1). State-of-the-art procedures are based on the user performing predefined static poses or functional movements, which make such a system less easy to use and can favor human-induced errors (cf. [Bou15]). An emerging field of research are self-calibration methods, which determine the calibration parameters from sensor measurements without prior knowledge or assumptions about the performed movements, e.g., [See14] proposes a method for two linked segments and [Tae16] proposes a promising proof-of-principle for an online-capable calibration correction and segment kinematics estimation method for the lower limbs.

2.3 Potential Applications

IMU based 3D kinematics estimation enables the reconstruction of individual movement patterns and biomechanically interpretable data (such as joint ranges of motion, trajectories of joints or other anatomical landmarks, segment orientations, spatiotemporal locomotion parameters) in-field. This ability is useful for different application areas. Some popular examples together with the parameters of interest are summarized in the following:

- Clinical movement analysis [Che16, Ble17]. In numerous areas of medicine, quantitative movement analysis, e.g., gait analysis, has proven effective in supporting assessments, diagnoses, and therapies. Here, a lightweight and easy-to-use wearable measurement system could support functional diagnostics and valid follow-up and documentation in everyday clinical practice.
- Rehabilitation exercises [Ble13, Lam15]. Exploiting the online processing capabilities of IMU based 3D kinematics estimation, this can also be used to promote and support self-training by providing direct feedback to patients on the movement quality. This is often combined with game-based features to increase motivation (e.g., [Gor17, Ste17]).
- Sports [Won15, Men16]. In-field capturing capabilities are of particular interest in the area of sports. Here, IMU based 3D kinematics estimation can be used e.g., for analyzing and improving athletic performance (cf. [Ree16] for an analysis of running kinematics during a marathon) or for investigating and treating sport-specific injuries (e.g., assessing leg axis stability after an anterior cruciate ligament injury [Che16]).
- Workflow analysis and assistance. In this area, IMU based 3D kinematics estimation can be used for different purposes, e.g., for: (1) designing and raising the awareness for ergonomically safe workflows (see [Vig13] for an IMU based system providing real-time ergonomic feedback on hazardous postures), (2) providing user monitoring as ingredient for building an intelligent workflow assistance system [Ble15a], (3) providing kinematic information to control wearable assistive devices, e.g., exoskeletons to support overhead work.

Other application areas for IMU based 3D kinematics estimation, which are only shortly mentioned here, are entertainment (e.g., gesture-based game control, animation of virtual characters) and robotics (e.g., teleoperation).

3 Human Activity Recognition Based on Multimodal Body Sensor Networks

3.1 Motivation

Moving away from the notion of having the exact trajectory of each body part as starting point for activity analysis is justified by three considerations. First, many activities have a distinct motion signature that can be detected even by a single sensor at various body locations. As an example, consider step detection. On the one hand the tracking of the trajectory of at least upper and lower legs is needed for exact step analysis. On the other hand, the up and down motion associated with each step and the shock of the foot hitting the ground produces a distinct acceleration signature that can be easily detected at nearly all body locations (this is how commercial step counters work). Second, for many activities there are important sources of information beyond the tracking of body parts kinematics. As an example, consider grasping an object and putting it on the table. The grasping motion of the fingers, including an estimation of the force, can be derived from the activity of the muscles in the lower arm which in terms can be sensed using electromyography (EMG), capacitive sensing, or textile pressure sensing matrices. In addition, putting the object on the table often produces a characteristic sound which can be detected with a body worn microphone. Given the above additional sensing modalities a rough estimate of the overall arm motion that can be derived from a single wrist worn IMU may be sufficient in terms of motion information. Third, there are many activities where body motions (at least in terms of limbs trajectories), are more or less irrelevant. A good example are cognition dominated activities such as reading, watching a movie, or having a conversation. In addition to obvious sources of information such as audio and first-person video, head motion patterns and eye tracking have been shown to be the key sources of information to distinguish such activities [Ish14].

3.2 Abstract Motion Signatures

As an example of an abstract motion signature the acceleration signal produced by a sensor in a trouser side pocket when the user is walking up and down stairs is shown in Fig. 3. A close inspection shows that there is obvious structure in the data which can be mapped onto features of human steps. We have one part corresponding to the leg being put forward and one for the leg being pulled from behind. In the walking downstairs case we have sharp peaks corresponding to the impact of the foot on the lower step. In the walking up signal soft peaks caused by the lifting and straightening of the leg can be seen. There are three main approaches for the automatic analysis of such signals.

First, abstract features such as mean, variance, root mean square (RMS), and frequency distribution can be computed on a sliding window with a length corresponding to the time scale of the underlying activity. In the case of step analysis this corresponds to around one second (typical step frequency). The features are then

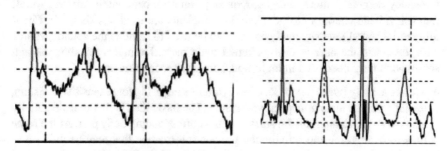

Fig. 3 Signals from an acceleration sensor worn by a person in a trouser side pocket while taking two steps up (left) and down (right) the stairs

fed into statistical classifiers such as Support Vector Machines (SVMs) or neural networks for the recognition of the underlying activities. For simple tasks such as modes of locomotion recognition (walking, running, walking up/down stairs etc.) such a simple approach is often sufficient.

Second, probabilistic time series modeling methods can be applied to better capture the temporal characteristics of the activity in question. Traditionally, Hidden Markov Models [Moj12, Dav16] (HMMs) have been widely used for wearable activity recognition. Related methods are Conditional Random Fields or various other variations of Dynamic Bayesian Networks (of which HMMs are just a special case). In most cases a separate model is trained for each activity. For recognition each of the trained models are applied to the signal and the one producing the highest probability is selected (which means the corresponding class is recognized). Such models have been used for the recognition of manipulative gestures (picking objects up, operating tools, eating, drinking etc.) from wrist/arm worn motion sensors, which are, in general, harder to separate than modes of locomotion [Jun08].

Third, template matching methods such as Dynamic Time Warping (DTW) can be applied to detect characteristic signal parts [Pha10]. To this end an "average" signal is computed from a large number of examples for each class. For recognition the class whose template is the best match is selected.

3.3 Combination of Abstract Motion and Other Information

When abstract motion signatures fail to provide sufficient discriminative power additional sensing modalities can help. For many activities sound is a very rich source of information. A good example is the use of tools such as a hammer, screwdriver, saw, drill etc. in a wood workshop task [War06]. The respective motions can be quite subtle and difficult to detect in a continuous stream of data. However, the activities have very distinct sounds associated with them. In general, frequency transformations on windows anywhere between 100 ms and 1 s followed by either linear discriminant analysis (LDA), principle component analysis (PCA) or computation of standard frequency domain features such as frequency centroid, bandwidth, spectral rolloff frequency, band energy ratio or cepstral coefficients are used. As shown in Fig. 4 differential sound intensity analysis can also be used to localize the sound's source with respect to the user or even different body parts. Examples of other relevant sensors that have been used in multimodal activity recognition are:

- Muscle activity [Ogr07, Amf06, Che12] monitoring using force sensitive resistors, textile pressure mats or capacitive sensors. The basic idea is that muscle activity leads to shape changes on the surface of the corresponding body part. At the same time looking at muscle activity can provide information that may be difficult to access using direct motion sensors such as IMUs. Thus, the motion of fingers and the palm is driven by muscles in the lower arm where sensors can be mounted much more easily than on the fingers themselves.

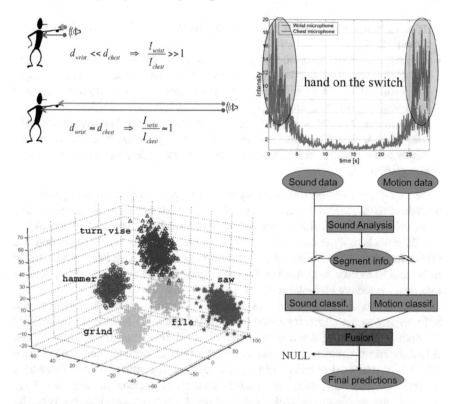

Fig. 4 Augmenting the recognition of wood workshop activities through sound processing [War06]. Top: using sound intensity difference between a microphone on the wrist and on the chest to identify sounds that originate close to the hand (e.g., from a machine that a hand is operating). Bottom left: Using linear discriminant analysis (LDA) of the frequency distribution to discriminate between the sounds of different tools. Bottom right: The overall recognition architecture

- Body sound. Many processes going on inside the human body create sounds that can be detected with appropriate wearable microphones. Examples range from muscle and joint motion through breathing, heartbeat and coughing to chewing and swallowing. Thus, for example an ear worn microphone can reliably detect chewing including the distinction between different types of food [Amf05].
- Hand tracking using ultrasonic tracking [Ogr12]. The idea is that the location of the hand with respect to the object on which the user is working is an important piece of information with respect to the user's activity. An alternative approach is to use a lower arm mounted radio-frequency identification (RFID) reader and RFID tags placed on objects.
- When working with IMUs magnetic fields and ferromagnetic objects in the environment are often seen as a problem as they disturb the signal (cf. Sect. 2.2). However, such disturbances can also be seen as a source of useful information. Thus, in general, different object appliances and machines will have a unique mag-

netic signature which can be used to recognize when the body worn IMU is near them [Bah10].

- Air pressure. While absolute air pressure depends on the weather and is not useful for activity recognition, fine air pressure variations can be used to detect changes in sensors' altitude corresponding to activities such as walking up or down stairs or even sitting down or standing up.
- Furthermore, high-level background information such as location, credit card transactions, data from autonomous devices such as smart home components, power consumptions in buildings and similar can be used to enhance activity recognition.

Since different sensors produce very different types of signals, feeding them into a single classifier seldomly produces good results. Instead, hierarchical, multi-stage recognition architectures have been exploited where different sensors are used to classify different activity components and the final recognition is done with appropriate sensor fusion (see Fig. 4). Recently, deep learning systems have been demonstrated to be able to replace such hand-crafted architectures with automatic extractions of intermediate feature hierarchies [Ord16].

Note, as already mentioned in Sect. 1, central differences of the described approach with respect to the above mentioned one (IMU based 3D kinematics estimation) are both in the level of detail with which information is reconstructed (recognized activities versus exact body motions) and the type of methods and models used (black-box statistical models and machine learning algorithms, which result in a dependence on training data, versus model-based sensor fusion methods, which are generally applicable and provide biomechanically interpretable data, but typically have stricter placement and attachment constraints).

4   Existing and Potential Synergies

While the two approaches (IMU based 3D kinematics estimation and general human activity recognition) have been presented separately in the above sections, there are indeed many existing and potential synergies both on method and on application level. Some of these are shortly indicated in the following.

- Sensor reduction. Consider the case where a full-body 3D kinematics reconstruction is needed, but the mounting of sensors on all body segments (e.g., 17 IMUs in commercially available systems [Xse17]) is infeasible. Here, large datasets of precisely captured motion (e.g., using a full IMU setup or a gold standard capturing system) have been combined with machine learning algorithms for reconstructing the full-body kinematics with a reduced amount of IMUs; e.g., Tautges et al. [Tau11] uses four accelerometers and Wouda et al. [Wou16] uses five IMUs. In [Mar17], visually pleasing results were obtained with six IMUs using an offline global optimization approach together with kinematic constraints. Obviously, such approaches introduce a dataset dependence and come at the cost of reduced accuracy, which can, however, be sufficient for specific applications.

- Automatic sensor assignment to body locations. Setting up a system with multiple IMUs can be error-prone regarding the correct placement of the IMUs on the different body segments. Here, machine learning algorithms [Kun05, Kun07, Kun14, Wee13, Zim18], as well as, hierarchical construction-based methods [Gra16] have been applied to obtain an automatic assignment during a predefined movement (e.g., walking). This has been used as pre-processing step for both IMU based 3D kinematics estimation and general activity recognition.

- Reduction of soft tissue and clothing artifacts. Soft tissue [Lea05] and clothing artifacts concern all body-mounted measurement systems. They constitute a major source of error in both IMU based 3D kinematics estimation and activity recognition [Moh17]. Recent literature provides initial model-based [Kok14] and data-level [Ols17] approaches to address soft tissue artifacts. Based on studies with optical markers attached to the skin [Cam12], Olsson and Halvorsen [Ols17] argue that a linear model is sufficient to compensate for soft tissue artifacts. In [Men15], soft tissue artifacts are compensated for in IMU based 3D kinematics estimation of the upper limbs via a linear regression approach that considers the person's body and arm total mass, fat mass, lean mass, and fat percentage. The regression is then used to obtain corrected estimates for planar arm movements. Integrating sensors into comfortable (i.e., not very tight) clothes (smart textiles) makes it more feasible to wear multiple sensors over a longer period of time. However, this also results in more severe artifacts. In [Moh17], a deep learning approach is proposed for increasing the signal-to-noise ratio for the case of IMUs being integrated into a training suit.

- Automatic segmentation of repetitive motions (e.g., rehabilitation exercises). In [Ble15b], a machine learning approach is used to segment the motion data obtained from IMU based 3D kinematics estimation for counting and evaluating single exercise repetitions. There are several advantages of performing the segmentation on the level of reconstructed 3D kinematic data instead of on raw sensor signals, e.g., obtaining biomechanically interpretable features and being more independent from the sensing hardware (cf. also [Ble15a]). A fusion with complementary sensors, such as pressure insoles or mobile force sensors, could enhance both biomechanical analysis (moving from kinematics to kinetics) and activity segmentation.

- Locomotion analysis. In locomotion analysis both kinematic (e.g., joint angles and segment orientations) and spatiotemporal parameters are of importance (cf. Sect. 2.3). While IMU based 3D kinematics estimation can immediately deliver the former, detection of the critical locomotion events (e.g., initial and terminal contact) is required for deducing the latter from the kinematics data. This is typically based on machine learning algorithms [Che16]. Note, in well-defined scenarios, such as walking straight on flat ground, spatiotemporal parameters have also solely been extracted based on machine learning algorithms, e.g., from two shoe-mounted IMUs using deep convolutional neural networks [Han17].

- Long-term context-sensitive biomechanical analysis, e.g., for the purpose of (medical) movement analysis in everyday life (instead of in a specific assessment situation) or for ergonomic feedback throughout the workday, could be a further scenario for a potential synergy. Here, classifications based on few sensors (to reduce energy consumption) could be used to trigger detailed biomechanical analyzes with additional sensors, only if relevant activities have been detected (e.g., normal walking, standing up/sitting down, or manipulating high weights).

5 Conclusion

Human motion capturing and activity recognition using wearable sensor networks represent enabling technologies, which can be used to enhance support/assistance systems in many application areas (e.g., in healthcare and sports, workflow analysis, human-computer-interaction, robotics, and entertainment, as already exemplified in the above sections). Knowledge about a user's motion, activity and possible environment allows assistance systems and assistive devices to adapt to the user and his or her context. This can improve usability and usefulness; e.g., think of Augmented Reality manuals which provide step-by-step guidance for manual workflows, exoskeletons to support overhead work which regulate the amount of support based on the wearer's motions and activities, leg prostheses which adapt their settings to best support the wearer's intended activity (e.g., standing up, climbing stairs). Another aspect concerns the improvement of human-machine-interaction; e.g., think of social robotics where knowledge about the human partners' motions or activities is essential to enable natural interactions. Enabling the provision of (online) feedback concerning a person's motion or activity is another aspect, which can be beneficial, e.g., think of motor learning in the context of rehabilitation or sports where feedback is both effective and motivational. An obvious aspect concerns the ability to provide in-field monitoring (either for online feedback or documentation), e.g., of daily living activities, specific movement patterns, but also sleep, nutrition or other body functions. These are all relevant in the health context, e.g., think of telemedicine, home based rehabilitation or early diagnosis. Another area is ergonomics, where in-field biomechanical analyzes can be helpful to design and raise the awareness of ergonomically safe workflows. In all of these real-world examples, respective support/assistance systems require or can at least benefit from mobile (i.e., location-independent instead of stationary hardware dependent) information gathering. This can be provided based on wearable sensor networks.

Acknowledgements This article is a joint work of the Interdisciplinary Junior Research Group wearHEALTH at TUK, funded by the BMBF (16SV7115), and the Embedded Intelligence Department at DFKI. Gabriele Bleser and Bertram Taetz (wearHEALTH) focused on IMU based 3D kinematics estimation, while Paul Lukowicz (Embedded Intelligence) focused on human activity recognition based on multimodal body sensor networks.

References

[Amf05]   Amft, O., Stager, M., Lukowicz, P., & Tröster, G. (2005). Analysis of chewing sounds for dietary monitoring. *UbiComp, 5,* 56–72.

[Amf06]   Amft, O., Junker, H., Lukowicz, P., Tröster, G., & Schuster, C. (2006). Sensing muscle activities with body-worn sensors. In *International Workshop on Wearable and Implantable Body Sensor Networks (BSN).* IEEE.

[Bah10]   Bahle, G., Kunze, K., & Lukowicz, P. (2010). On the use of magnetic field disturbances as features for activity recognition with on body sensors. In *European Conference on Smart Sensing and Context* (pp. 71–81). Berlin, Heidelberg: Springer.

[Bao04]   Bao, L., & Intille, S. (2004). Activity recognition from user-annotated acceleration data. In *Pervasive Computing* (pp. 1–17).

[Ble09]   Bleser, G., & Stricker, D. (2009). Advanced tracking through efficient image processing and visual-inertial sensor fusion. *Computer & Graphics, 33,* 59–72.

[Ble13]   Bleser, G., Steffen, D., Weber, M., Hendeby, G., Stricker, D., Fradet, L., et al. (2013). A personalized exercise trainer for the elderly. *Journal of Ambient Intelligence and Smart Environments, 5,* 547–562.

[Ble15a]   Bleser, G., Damen, D., Behera, A., Hendeby, G., Mura, K., Miezal, M., et al. (2015). Cognitive learning, monitoring and assistance of industrial workflows using egocentric sensor networks. *PLoS ONE, 10*(6), e0127769.

[Ble15b]   Bleser, G., Steffen, D., Reiss, A., Weber, M., Hendeby, G., & Fradet, L. (2015). Personalized physical activity monitoring using wearable sensors. In *Smart health* (pp. 99–124). Cham: Springer.

[Ble17]   Bleser, G., Taetz, B., Miezal, M., Christmann, C. A., Steffen, D., & Regenspurger, K. (2017). Development of an inertial motion capture system for clinical application—Potentials and challenges from the technology and application perspectives. *Journal of Interactive Media, 16*(2).

[Bou15]   Bouvier, B., Duprey, S., Claudon, L., Dumas, R., & Savescu, A. (2015). Upper limb kinematics using inertial and magnetic sensors: Comparison of sensor-to-segment calibrations. *Sensors, 15*(8), 18813–18833.

[Cam12]   Camomilla, V., Cereatti, A., Cheze, L., Cappozzo, A. (2012). A hip joint kinematics driven model for the generation of realistic thigh soft tissue artefacts. *Journal of Biomechanics, 46*(3), 625–630.

[Cam16]   Chen, S., Lach, J., Lo, B., & Yang, G.-Z. (2016). Towards pervasive gait analysis for medicine with wearable sensors: A systematic review for clinicians and medical researchers. *IEEE Journal of Biomedical and Health Informatics, 20*(6), 1521–1537.

[Che12]   Cheng, J., Bahle, G., & Lukowicz, P. (2012). A simple wristband based on capacitive sensors for recognition of complex hand motions. *IEEE Sensors Journal,* 1–4.

[Dav16]   Davis, K., Owusu, E., Bastani, V., Marcenaro, L., Hu, J., Regazzoni, C., et al. (2016). Activity recognition based on inertial sensors for ambient assisted living. In *19th International Conference on Information Fusion (FUSION)* (pp. 371–378).

[Fas17]   Fasel, B., Spörri, J., Schütz, P., Lorenzetti, S., & Aminian, K. (2017). Validation of functional calibration and strapdown joint drift correction for computing 3D joint angles of knee, hip, and trunk in alpine skiing. *PLoS ONE, 12*(7).

[Gor17]   Gordt, K., Gerhardy, T., Najafi, B., & Schwenk, M. (2017). Effects of wearable sensor-based balance and gait training on balance, gait, and functional performance in healthy and patient populations: A systematic review and meta-analysis of randomized controlled trials. *Gerontology.*

[Gra16]   Graurock, D., Schauer, T., & Seel, T. (2016). Automatic pairing of inertial sensors to lower limb segments—A plug-and-play approach. *Current Directions in Biomedical Engineering, 2*(1), 715–718.

[Gus12]   Gustafsson, F. (2012). *Statistical sensor fusion* (2nd edn.). Studentlitteratur.

[Han17]   Hannink, J., Kautz, T., Pasluosta, C. F., Gamann, K., Klucken, J., & Eskofier, B. M. (2017). Sensor-based gait parameter extraction with deep convolutional neural networks. *IEEE Journal of Biomedical and Health Informatics, 21*(1), 85–93.

[Har13]   Harle, R. (2013). A survey of indoor inertial positioning systems for pedestrians. *IEEE Communications Surveys & Tutorials, 15,* 1281–1293.

[Hol11]   Hol, J. D. (2011). *Sensor fusion and calibration of inertial sensors, vision, ultra-wideband and GPS.* Ph.D. thesis, Linkping University, Department of Electrical Engineering, Automatic Control.

[Ios16]   Iosa, M., Picerno, P., Paolucci, S., & Morone, G. (2016). Wearable inertial sensors for human movement analysis. *Expert Review of Medical Devices.*

[Ish14]   Ishimaru, S., Kunze, K., Kise, K., Weppner, J., Dengel, A., Lukowicz, P., et al. (2014). In the blink of an eye: Combining head motion and eye blink frequency for activity recognition with google glass. In *Proceedings of the 5th Augmented Human International Conference* (p. 15). ACM.

[Jun08]   Junker, H., Amft, O., Lukowicz, P., & Tröster, G. (2008). Gesture spotting with bodyworn inertial sensors to detect user activities. *Pattern Recognition, 41*(6), 2010–2024.

[Kok14]   Kok, M., Hol, J., & Schön, T. (2014). An optimization-based approach to human body motion capture using inertial sensors. In *Proceedings of the 19th World Congress of the International Federation of Automatic Control (IFAC)* (pp. 79–85).

[Kok15]   Kok, M., Hol, J. D., & Schön, T. B. (2015). Indoor positioning using ultrawideband and inertial measurements. *IEEE Transactions on Vehicular Technology, 4*(64), 1293–1303.

[Kok16]   Kok, M., Hol, J. D., & Schön, T. B. (2016). *Using inertial sensors for position and orientation estimation.*

[Kun05]   Kunze, K., Lukowicz, P., Junker, H., & Tröster, G. (2005). Where am I: Recognizing on-body positions of wearable sensors. In *International Symposium on Location-and Context-Awareness* (pp. 264–275). Springer.

[Kun07]   Kunze, K., & Lukowicz, P. (2007). Using acceleration signatures from everyday activities for onbody device location. In *11th IEEE International Symposium on Wearable Computers* (pp. 115–116).

[Kun14]   Kunze, K., & Lukowicz, P. (2014). Sensor placement variations in wearable activity recognition. *IEEE Pervasive Computing, 13*(4), 32–41.

[Lam15]   Lam, A. W. K., Varona-Marin, D., Li, Y., Fergenbaum, M., & Kulic, D. (2015). Automated rehabilitation system: Movement measurement and feedback for patients and physiotherapists in the rehabilitation clinic. *Human Computer Interaction, 31,* 294–334.

[Lea05]   Leardini, A., Chiari, L., Croce, U. D., & Cappozzo, A. (2005). Human movement analysis using stereophotogrammetry: Part 3. soft tissue artifact assessment and compensation. *Gait & Posture, 21*(2), 212–225.

[Lea07]   Leardini, A., Sawacha, Z., Paolini, G., Ingrosso, S., Nativo, R., & Benedetti, M. G. (2007). A new anatomically based protocol for gait analysis in children. *Gait & Posture, 26,* 560–571.

[Lig16]   Ligorio, G., & Sabatini, A. M. (2016). Dealing with magnetic disturbances in human motion capture: A survey of techniques. *Micromachines, 7*(3).

[Lop16]   Lopez-Nava, I. H., & Angelica, M.-M. (2016). Wearable inertial sensors for human motion analysis: A review. *IEEE Sensors Journal.*

[Mar17]   von Marcard, T., Rosenhahn, B., Black, M. J., & Pons-Moll, G. (2017). Sparse inertial poser: Automatic 3D human pose estimation from sparse imus. *Computer Graphics Forum, 36,* 349–360.

[Men15]   Meng, D., Shoepe, T., & Vejarano, G. (2015). Accuracy improvement on the measurements of human-joint angles. *IEEE Journal of Biomedical and Health Informatics, 2*(20), 498–507.

[Men16]   Mendes, J. J. A, Vieira, M. E. M., Pires, M. B., & Stevan, S. L. (2016). Sensor fusion and smart sensor in sports and biomedical applications. *MDPI Sensors, 16*(10).

[Mie16]   Miezal, M., Taetz, B., & Bleser, G. (2016). On inertial body tracking in the presence of model calibration errors. *MDPI Sensors, 16*(7).

[Mie17]   Miezal, M., Taetz, B., & Bleser, G. (2017). Real-time inertial lower body kinematics and ground contact estimation at anatomical foot points for agile human locomotion. In *International Conference on Robotics and Automation*, Singapore.
[Moh17]   Mohammed, S., & Tashev, I. (2017). Unsupervised deep representation learning to remove motion artifacts in free-mode body sensor networks. In *14th International Conference on Wearable and Implantable Body Sensor Networks (BSN)* (pp. 183–188).
[Moj12]   Mojidra, H. S., & Borisagar, V. H. (2012). A literature survey on human activity recognition via hidden markov model. In *IJCA Proceedings on International Conference on Recent Trends in Information Technology and Computer Science*.
[Ogr07]   Ogris, G., Kreil, M., & Lukowicz, P. (2007). Using FSR based muscle activity monitoring to recognize manipulative arm gestures. In *11th International Symposium on Wearable Computers* (pp. 45–48). IEEE.
[Ogr12]   Ogris, G., Lukowicz, P., Stiefmeier, T., & Tröster, G. (2012). Continuous activity recognition in a maintenance scenario: Combining motion sensors and ultrasonic hands tracking. *Pattern Analysis and Applications, 15*(1), 87–111.
[Ols17]   Olsson, F., & Halvorsen, K. (2017). Experimental evaluation of joint position estimation using inertial sensors. In *20th International Conference on Information Fusion (Fusion)* (pp. 1–8). IEEE.
[Ord16]   Ordóñez, F. J., & Roggen, D. (2016). Deep convolutional and LSTM recurrent neural networks for multimodal wearable activity recognition. *MDPI Sensors, 16*(1), 115.
[Pap17]   Papi, E., Koh, W. S., & McGregor, A. H. (2017). Wearable technology for spine movement assessment: A systematic review. *Journal of Biomechanics*.
[Pha10]   Pham, C., Plötz, T., & Olivier, P. (2010). A dynamic time warping approach to realtime activity recognition for food preparation. In *Ambient Intelligence* (pp. 21–30).
[Pic17]   Picerno, P. (2017). 25 years of lower limb joint kinematics by using inertial and magnetic sensors: A review of methodological approaches. *Gait & Posture, 51*, 239–246.
[Ree16]   Reenalda, J., Maartens, E., Homan, L., & Jaap Buurke, J. H. (2016). Continuous three dimensional analysis of running mechanics during a marathon by means of inertial magnetic measurement units to objectify changes in running mechanics. *Journal of Biomechanics*.
[Sch99]   Schmidt, A., Aidoo, K. A., Takaluoma, A., Tuomela, U., Van Laerhoven, K., & Van de Velde, W. (1999). Advanced interaction in context. In *HUC* (Vol 99, pp. 89–101). Berlin, Heidelberg: Springer.
[See14]   Seel, T., Schauer, T., & Raisch, J. (2014). IMU-based joint angle measurement for gait analysis. *MDPI Sensors, 14*(4), 6891–6909.
[Shu14]   Shull, P. B., Jirattigalachote, W., Hunt, M. A., Cutkosky, M. R., & Delp, S. L. (2014). Quantified self and human movement: A review on the clinical impact of wearable sensing and feedback for gait analysis and intervention. *Gait & Posture*.
[Ste17]   Steffen, D., Christmann, C. A., & Bleser, G. (2017) jumpball - ein mobiles exergame für die Thromboseprophylaxe. In *Mensch und Computer*.
[Sut02]   Sutherland, D. H. (2002). The evolution of clinical gait analysis: Part ii kinematics. *Gait & Posture, 16*(2), 159–179.
[Tae16]   Taetz, B., Bleser, G., & Miezal, M. (2016). Towards self-calibrating inertial body motion capture. In *19th International Conference on Information Fusion* (pp. 1751–1759). IEEE.
[Tag14]   Tagliamonte, N. L., Peruzzi, A., Accoto, D., Cereatti, A., Della Croce, U., & Guglielmelli, E. (2014). Assessment of lower limbs kinematics during human–robot interaction using inertial measurement units. Gait & Posture, 40.
[Tau11]   Tautges, J., Zinke, A., Krüger, B., Baumann, J., Weber, A., Helten, T., et al. (2011). Motion reconstruction using sparse accelerometer data. *ACM Transactions on Graphics, 30*(3), 18, 1–18, 12.
[Thr05]   Thrun, S., Burgard, W., & Fox, D. (2005). *Probabilistic robotics (Intelligent robotics and autonomous agents)*. Cambridge: The MIT Press.
[Vig13]   Vignais, N., Miezal, M., Bleser, G., Mura, K., Gorecky, D., & Marin, F. (2013). Innovative system for real-time ergonomic feedback in industrial manufacturing. *Applied Ergonomics, 44*(4), 566–574.

[War06] Ward, J. A., Lukowicz, P., Tröster, G., & Starner, T. E. (2006). Activity recognition of assembly tasks using body-worn microphones and accelerometers. *IEEE Transactions on Pattern Analysis and Machine Intelligence, 28*(10), 1553–1567.

[Wee13] Weenk, D., Van Beijnum, B.-J. F., Baten, C. T. M., Hermens, H. J., & Veltink, P. H. (2013). Automatic identification of inertial sensor placement on human body segments during walking. *Journal of Neuroengineering and Rehabilitation, 10*(1).

[Wen15] Wenk, F., & Frese, U. (2015). Posture from motion. In *International Conference on Intelligent Robots and Systems (IROS)* (pp. 280–285). IEEE.

[Won15] Wong, C., Zhang, Z.-Q., Lo, B., & Yang, G.-Z. (2015). Wearable sensing for solid biomechanics: A review. *IEEE Sensors Journal, 15*(5), 2747–2760.

[Wou16] Wouda, F. J., Giuberti, M., Bellusci, G., & Veltink, P. H. (2016). Estimation of full-body poses using only five inertial sensors: An eager or lazy learning approach? *MDPI Sensors*.

[Xse17] Xsens Technologies B.V. Xsens mvn website: https://www.xsens.com/products/xsens-mvn/. November 2017.

[Zhe14] Zheng, Y., Ding, X., Poon, C., Lo, B., Zhang, H., Zhou, X., et al. (2014). Unobtrusive sensing and wearable devices for health informatics. *IEEE Transactions on Biomedical Engineering, 61*(5), 1538–1554.

[Zim18] Zimmermann, T., Taetz, B. & Bleser, G. (2018) IMU-to-segment assignment and orientation alignment for the lower body using deep learning. *Sensors (Basel), 18*(1).

Soft Robotics. Bio-inspired Antagonistic Stiffening

Agostino Stilli, Kaspar Althoefer and Helge A. Wurdemann

Abstract Soft robotic structures might play a major role in the 4th industrial revolution. Researchers have demonstrated advantages of soft robotics over traditional robots made of rigid links and joints in several application areas including manufacturing, healthcare, and surgical interventions. However, soft robots have limited ability to exert larger forces and change their stiffness on demand over a wide range. Stiffness can be achieved as a result of the equilibrium of an active and a passive reaction force or of two active forces antagonistically collaborating. This paper presents a novel design paradigm for a fabric-based Variable Stiffness System including potential applications.

1 Introduction

With the growing interest in the use of elastic materials for the creation of highly dexterous robots [Bau14], material science has made inroads in the soft robotics community and becomes of paramount importance when creating soft robotic structures. A clear indication is the growth of publications about innovative soft material robots in recently appearing monothematic journals such as "Soft Robotics" [Rob99] and dedicated sessions at major robotics conferences, e.g., ICRA and IROS. Some roboticists argue that soft robotic technologies will play a key role in the 4th industrial revolution [Ros16], for safe human-robot interaction in manufacturing

A. Stilli
Department of Computer Science, University College London, London WC1E 7JE, UK
e-mail: a.stilli@ucl.ac.uk

K. Althoefer
School of Engineering and Materials Science, Queen Mary University of London, London E1 4NS, UK

H. A. Wurdemann (✉)
Department of Mechanical Engineering, University College London, London WC1E 7JE, UK
e-mail: h.wurdemann@ucl.ac.uk
URL: http://softhaptics.website

© Springer Nature Switzerland AG 2018   207
A. Karafillidis and R. Weidner (eds.), *Developing Support Technologies*, Biosystems & Biorobotics 23, https://doi.org/10.1007/978-3-030-01836-8_20

[Sti16, Sti17, Pfe13], healthcare [Hor17], and minimally invasive surgery (MIS) [Are16]. Numerous proposals for novel flexible robots based on soft and hybrid materials are continuously emerging [Lip14]. Continuum hyper-redundant designs have been extensively investigated to create soft robots for applications in surgery [Bur15] with embedded sensors [Far14, Sar14, Wur15a], in disaster scenarios [Kam04], and for underwater exploration [Cre05].

Although recent advances in soft and soft material robotics are notable and hold considerable promise to achieve what has not been possible with traditional rigid-linked robots, one important drawback remains: Despite their morphological capabilities, they have limited ability to exert larger forces on the environment when required, i.e., to change their stiffness on demand over a wide range [Man16]. In the search for the right trade-off between desired compliance and exertable force, researchers explored numerous approaches to enable on-demand stiffness tuning of soft robots. According to the recent comparative study presented in [Man16], Variable Stiffness Systems (VSSs) for soft robots can be divided in two main groups: (i) Active VSSs (AVSSs): these VSSs provide on-demand stiffening using an antagonistic approach, i.e., the creation of stiffness by means of equilibrium between two or more forces, at least one of which is an active force and (ii) Semi-Active VSSs (SAVSSs): these VSSs provide on demand stiffening relying on their capability of intrinsically tuning the rigidity of the robotic system in which they are embedded. In particular, recent works in [Mag15, Sti14, Shi16, Wur15b, Haw17] open up new avenues in this area by proposing inflatable robotic devices for applications in difficult-to-access sites, as in MIS and remote inspection. These devices can be highly compacted when in their undeployed, folded state and can be expanded in volume to multiples of the folded-state volume by injecting fluid and changing its stiffness by multiple times.

Our paper presents Variable Stiffness Systems based on a novel design paradigm for fabric-based soft robots. These stiffness mechanisms are inspired by nature and based on an active and passive antagonistic actuation principle. We describe the generic design, fabrication process, and capabilities of these robotic systems including potential application areas that have been explored.

2   Bio-inspired Embodiment of a Stiffness Mechanism

The stiffness mechanism proposed in this paper is inspired by biology: The role model for our research is the arm of the octopus. An octopus' arms are made of longitudinal and transversal muscles [Coa11]. Activating these muscle pairs that are distributed along the arms, the octopus is able to achieve high stiffness values. In other words, the sets of muscles "collaborate" in an opposing way to antagonistically stiffen the entire arm or arm segments. Hence, the octopus' arms are muscular hydrostats and can alternate between soft and rigid states combining advantages associated with both soft and hard systems by selectively controlling the stiffness of various parts of the body depending on the task requirements. Scientists have identified connecting tissue that keeps the muscles of the octopus arms in place, avoiding bulging and allowing the animal to achieve stiffness in their arms (comparable to a tube inside a bicycle tire). Our proposed actuation approach is antagonistic in its nature in the same way. To achieve similar behavior, we have combined compliant (silicone and rubber materials) and non-compliant (fabric meshes and textiles) materials with the passive and active antagonistic manipulation principles.

3 The Variable Stiffness Link: A Novel Active Structural Element

Silicone-based structures have been widely explored. They aim at imitating biological behavior and achieving robotic solutions for complex challenges. The morphology of these robotic structures has been exploited creating adaptable systems capable of inherently safe human-robot and environment-robot interaction. The main idea is to fabricate chambers of complex shapes with embedded fluidic actuators creating active variable stiffness systems, both, in active-active and active-passive configurations. To achieve a wide range of force and stiffness variation, braided material has been integrated to provide additional structural constraints to the chambers, thus, limiting undesired deformations—also known as ballooning. However, spacing between threads forming the braiding occurs and increases at high pressure values resulting in limitations.

3.1 Design Paradigm and Methodology

To prevent the aforementioned ballooning phenomena, we propose a novel design paradigm: the use of fabric as external braiding for fluidically actuated soft robotic structures. This approach has been firstly implemented in the creation of a novel active structural element called the Variable Stiffness Link (VSL), an AVSS working in Acitve-Passive configuration. The proposed system is shown in Fig. 1a and comprises a hollow cylindrical structure made of silicone material, an embedded plastic mesh, and an external nylon fabric sleeve. The internal airtight cylindrical chamber is supplied with pressurized air controlling the stiffness of the system. On the one hand, the plastic mesh guarantees shape retention at low pressures, on the other hand, the fabric layer provides a robust shape constraint at high pressures preventing any undesired deformation. The balance between the internal air pressure and the reaction force by the fabric sleeve defines the stiffness of the VSL. Interfacing the VSL with a pressure regulator which monitors the current internal pressure allows to detect rapid changes in pressure values resulting from physical interaction with the environment.

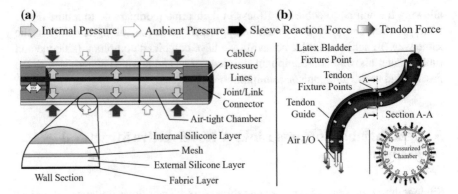

Fig. 1. Cross-sectional view of **a** the Variable Stiffness Link made of a plastic mesh embedded into silicone material and **b** the Inflatable Arm made of a latex bladder inside a fabric sleeve

3.2 Materials and Fabrication

The fabrication process of the VSL is as follows: a rectangular sheet of polypropylene diamond-shaped mesh is closed in a cylindrical shape by sealing the two overlapping edges at low temperature. The overlap is kept to a minimum in order to minimize the thickness increase after sealing thus keeping the system isomorphic. By using a commercially available heat sealer, a sealing line is produced on the rolled-up rectangular mesh forming the mesh into the shape of a cylinder. During the second stage of the fabrication process, the plastic mesh is embedded into a layer of silicone. A two-phase molding process is applied to cast silicone material on the mesh into a cylindrical shape. The result is a light-weight cylindrical-shaped stiffness-controllable element with a large internal lumen.

4 Bio-inspired Actuation for a Soft Continuum Manipulator

Built on the pneumatic actuation of the VSL, we have further explored robotic structures that are able to change its stiffness as well as shape. To enable shape shifting and shape locking capabilities, an additional actuation means has been introduced. Taking inspiration from the stiffening mechanism of natural muscles, we have developed a novel design that makes use of the extension behavior of pneumatic actuators and the contraction behavior of tendon-driven actuators resulting in an active variable stiffness system with active-active configuration. This concept has been firstly implemented in a continuum robotic manipulator called the Inflatable Arm, an AVSS working in Active-Active configuration.

The actuation principle and the design of the Inflatable Arm are illustrated in Fig. 1b. The proposed system comprises three main elements: an internal airtight, yet expandable, latex bladder; an external, non-expandable, but collapsible and foldable polyester sleeve; and nylon tendons that are mounted to the outer fabric sleeve. Three tendons are fixed at the manipulator's tip and another set of three is attached to the outer fabric halfway between base and tip, 120° spaced apart along the perimeter of the outer sleeve. The pushing force of the pressurized air inflates the manipulator and provides a straitening momentum, while the pulling force of the tendons steer the manipulator in the desired direction. A stable equilibrium between these forces can be achieved in any configuration providing the desired stiffness-controllability, shape-shifting, and shape-locking capabilities. To show the potential of this approach in different application areas, this design paradigm has been implemented in two robotic systems, the Inflatable Endoscope and the Inflatable Exoskeleton Glove (INFLEX-OGlove).

5 Potential Applications for Industrial Settings and Healthcare

Three systems have been created to demonstrate the successful application of the presented robotic structures with embedded bio-inspired antagonistic stiffening:

- Figure 2a illustrates the concept for a collaborative robot made of VSLs. The idea is to replace the rigid links of serial robots with VSLs. Hence, it is possible to change the stiffness of the links by varying the value of pressure inside their structure. Moreover, pressure readings from the pressure sensors inside the regulators can be utilized to detect collisions between the manipulator body and a human worker, for instance.
- Figure 2b shows an inflatable, stiffness-controllable endoscope for minimally invasive surgery. Due to the nature of the used outer material and its soft, compressible

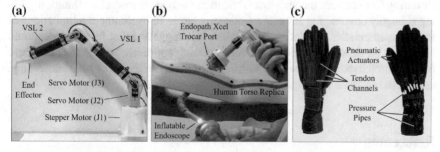

Fig. 2. Applications of the presented novel design paradigm: **a** a cobot made of Variable Stiffness Links, **b** an inflatable and stiffness-controllable endoscope and **c** the INFLatable EXOskeleton Glove (InflExoGlove)

structure, the proposed device is inherently safe when physically interacting with soft tissue.

- Figure 2c proposes a light-weight infltable soft exoskeleton device, called the InflExoGlove, to deliver gradual rehabilitation therapy and, hence, effective high-dosage rehabilitation therapies for post-stroke disabilities.

6 Conclusion

We presented a novel design paradigm of a stiffness concept that has been implemented in the creation of the VSL, an Active VSSs in Active-Passive configuration: The active part is the pneumatic actuation whereas the passive part is a structure based on flexible membranes in combination with fabric. The VSL is a stiffness-controllable structural element that can be embedded in soft robots or be used to replace rigid components in conventional robots. A further development of this concept is realised by the presented novel bio-inspired design for a soft continuum manipulator, the Inflatable Arm. This robotic system is able to control its stiffness on demand and also is endowed with shape-shifting and shape-locking capabilities. The Inflatable Arm is an Active VSS that works in Active-Active antagonistic configuration, where the first active mean is the pneumatic actuation and the second one is the tendon-driven actuation.

The fabrication processes presented in this paper have shown that the proposed soft, stiffness-controllable structures can be easily developed across different scales allowing applications in minimally invasive surgery as well as applications where these robots are utilised to work closely together with humans in industrial settings. These prototypes effectively undergo a crafting process combining multiple materials such as fabric meshes and silicone materials inside moulds in a number of procedural steps to finally create a new robotic structure. Future work will investigate how to industrialize the fabrication steps, by means of sewing machines and multi-material/multi-stiffness 3D printing, to allow precise reproducibility, in particular, when these soft, stiffness-controllable robots are working in interaction with humans. Resulting from current inconsistent fabrication methods among other things, modeling is hugely challenging for soft robotic structures. To overcome the limitations of today's mathematical modeling approaches, it is suggested that future work should also focus on deep learning and machine learning approaches—methods that have become attractive for applications with complex models affected by high uncertainty, such as the systems investigated in this paper.

References

[Are16]   Arezzo, A., Mintz, Y., Allaix, M. E., Arolfo, S., Bonino, M., Gerboni, G., et al. (2016). Total mesorectal excision using a soft and flexible robotic arm: A feasibility study in cadaver models. *Surgical Endoscopy, 31*(1), 264273.

[Bau14]   Bauer, S., Bauer-Gogonea, S., Graz, I., Kaltenbrunner, M., Keplinger, C., & Schwoedi-auer, R. (2014). 25th anniversary article: A soft future: From robots and sensor skin to energy harvesters. *Advanced Materials, 26*(1), 149162.

[Bur15]   Burgner-Kahrs, J., Rucker, D. C., & Choset, H. (2015). Continuum robots for medical applications: A survey. *IEEE Transactions on Robotics, 31*(6), 1261–1280.

[Coa11]   Cianchetti, M., Arienti, A., Follador, M., Mazzolai, B., Dario, P., & Laschi, C. (2011). Design concept and validation of a robotic arm inspired by the octopus. *Materials Science and Engineering C, 31,* 1230–1239.

[Cre05]   Crespi, A., Badertscher, A., Guignard, A., & Ijspeert, A. J. (2005). AmphiBot I: An amphibious snake-like robot. *Robotics and Autonomous Systems, 50*(4), 163–175.

[Far14]   Faragasso, A., Stilli, A., Bimbo, J., Noh, Y., Liu, H., et al. (2014). Endoscopic add-on stiffness probe for real-time soft surface characterisation in MIS. In: *International Conference of the IEEE Engineering in Medicine and Biology Society* (pp. 6517–6520).

[Haw17]   Hawkes, E. W., Blumenschein, L. H., Greer, J. D., & Okamura, A. M. (2017). A soft robot that navigates its environment through growth. *Science Robotics, 2*(8).

[Hor17]   Horvath, M. A., Wamala, I., Rytkin, E., Doyle, E., Payne, C. J., Thalhofer, T., et al. (2017). An intracardiac soft robotic device for augmentation of blood ejection from the failing right ventricle. *Annals of Biomedical Engineering, 45*(9), 2222–2233.

[Kam04]   Kamegawa, T., Yarnasaki, T., Igarashi, H., & Matsuno, F. (2004). Development of the snake-like rescue robot kohga. In *IEEE International Conference on Robotics and Automation* (pp. 5081–5086).

[Lip14]   Lipson, H. (2014). Challenges and opportunities for design, simulation, and fabrication of soft robots. *Soft Robotics, 1*(1), 21–27.

[Mag15]   Maghooa, F., Stilli, A., Noh, Y., Althoefer, K., & Wurdemann, H. A. (2015). Tendon and pressure actuation for a bio-inspired manipulator based on an antagonistic principle. In *IEEE International Conference on Robotics and Automation* (pp. 2556–2561).

[Man16]   Manti, M., Cacucciolo, V., & Cianchetti, M. (2016). Stiffening in soft robotics: A review of the state of the art. *IEEE Robotics and Automation Magazine, 23*(3), 93–106.

[Pfe13]   Pfeifer, R., Marques, H. G., & Iida, F. (2013). Soft robotics: The next generation of intel-ligent machines. In *International Joint Conference on Artificial Intelligence* (pp. 5–11).

[Rob99]   Robinson, G., & Davies, J. B. C. (1999). Continuum robots—A state of the art. In *IEEE International Conference on Robotics and Automation* (p. 4).

[Ros16]   Rossiter, J., & Hauser, H. (2016). Soft robotics the next industrial revolution. *IEEE Robotics and Automation Magazine, 23,* 17–20.

[Sar14]   Sareh, S., Jiang, A., Faragasso, A., Nanayakkara, T., Dasgupta, P., Seneviratne, L., Wurdemann, H., et al. (2014). MR-compatible bio-inspired tactile sensor sleeve for sur-gical soft manipulators. In *IEEE International Conference on Robotics and Automation* (pp. 1454–1459).

[Shi16]   Shiva, A., Stilli, A., Noh, Y., Faragasso, A., Althoefer, K., & Wurdemann, H. A. (2016). Tendon-based stiffening for a pneumatically actuated soft manipulator. *IEEE Robotics and Automation Letters, 1*(2), 632–637.

[Sti14]   Stilli, A., Wurdemann, H. A., & Althoefer, K. (2014). Shrinkable, stiffness-controllable soft manipulator based on a bio-inspired antagonistic actuation principle. In *IEEE Inter-national Conference on Robotics and Automation* (pp. 2476–2481).

[Sti16]   Stilli, A., Wurdemann, H. A., & Althoefer, K. (2016). A novel concept for safe. *Stiffness-Controllable Robot Links, Soft Robot, 4*(1), 16–22.

[Sti17]   Stilli, A., Grattarola, L., Feldmann, H., Wurdemann, H. A., & Althoefer, K. (2017). Variable Stiffness Links VSL—Toward inherently safe robotic manipulators. In *IEEE International Conference on Robotics and Automation* (pp. 4971–4976).

[Wur15a]   Wurdemann, H. A., Sareh, S., Shafti, A., Noh, Y., Faragasso, A., Chathuranga, D. S. (2015). Embedded electro-conductive yarn for shape sensing of soft robotic manipu-lators. In *International Conference of the IEEE Engineering in Medicine and Biology Society* (pp. 8026–8029).

[Wur15b]   Wurdemann, H., Stilli, A., & Althoefer, K. (2015). Lecture notes in computer science:
           An antagonistic actuation technique for simultaneous stiffness and position control. In
           International Conference on Intelligent Robotics and Applications (pp. 164–174).

Part IV
Values and Valuation

Assessment and evaluation of technical systems are necessary and common in technology development. But when devices, apparatuses, and gadgets are developed with a supporting purpose, the understanding of evaluation is carried far beyond a simple testing of functionalities and user satisfaction. Classic evaluation regarding operational functionality and safety issues is still inevitable, but it is complemented by further forms of evaluation. Technical systems get "worth" in a continuous process of valuation. Thereby, they gain (or lose) value in many respects, for example, ethical value, legal value, shared sociocultural value, monetary value, or functional value.

Values and valuation are an integral part of developing support technologies. Human beings dwell in a socio-techno-cultural web of relations. Developing for people and their desires (as well as, to be sure, invoking new desires and needs) ensues that any ignorance of specific values may quickly result in a failure—even if the technical systems seem promising from a purely functional point of view. This support system approach advocated here transcends the typical Ethical, Legal, and Social Implications (ELSI)" claims. ELSI no doubt paved the way but exhibits two major drawbacks: it is installed mainly outside the development processes, and it often comes after major technical choices have already been made. In contrast, the support systems approach subscribes itself more to the idea of "responsible research and innovation": Values and valuation are not part of an ex post assessment but imprinted into a project. Support systems define a kind of new paradigm for a reinterpretation of issues commonly termed "human–machine interaction".

Ongoing processes of valuation "in-form" the development continuously—in the same way demands analysis, construction, and deployment remain important during the whole course of a development project. This aspect of developing support technologies includes the fact that technological choices are also driven by societal values, even when they seem to be determined by technical considerations only. This has partly been discussed earlier in Part II of this volume on "Constructing and Construing". However, societal values are not homogeneous and consented. Society is differentiated into many network domains, modes of existence, linked ecologies, and subsystems. Its unity is a product of difference, including a difference of conflicting values, orders of worth, and the resulting friction. This might turn out as a problem in certain respects, yet it is also known to be the principal source of innovation.

This Part IV of the book starts with a paper of *Jörg Miehling, Alexander Wolf* and *Sandro Wartzack* who present how an evaluation of support system designs can be accomplished by building models of the human musculoskeletal structure to run simulations. Since it is impossible to continuously have test persons in the laboratory or to do field tests, having a proper model may save time and costs and therefore accelerates the process and also encourages organizations to try out support systems. Gaining knowledge about possible factors for success and failure of a system at each point in time by simulation is promising for determining the prospective value of a technical solution even in the early phases of its implementation.

Oliver Schürer, Christoph Müller, Christoph Hubatschke and *Benjamin Stangl* introduce the sketches of a model for an "intra-action" between humans and robots. Entering a common lifeworld, the authors argue, requires the creation of a similar understanding and perception of local space. Thus, an "intra"-action rather than "inter"-action unfolds that serves as a prerequisite for both entities to be able to value each other. The imputation of a shared cultural space provides a condition to evaluate the performance of the robot or any artificial intelligence application adequately.

Simulation is again the focus of *Paul Glogowski, Kai Lemmerz, Alfred Hypki,* and *Bernd Kuhlenkötter*. But this time the sociotechnical relation is approached from the opposite angle. A model for robot movements is presented that prepares the ground for an evaluation of a collaboration between humans and robots. The authors show how a widely used software framework for robots can be applied to model and visualize a robot's kinematics and dynamics that are relevant for collaboration with humans.

The first three chapters discuss relations of two basically separated entities, that is, robots on the one hand and human beings on the other. In the subsequent chapter, an evaluation of integrated human/technical systems is presented. *Christina M. Hein* and *Tim C. Lueth* discuss research designs for wearable aids to ascertain user acceptance. First, they give an overview of different types of tests and measurement tools for wearables (e.g., exoskeletons) and then go on to describe a field study in a nursing home. They compare two completely flexible wearable aids in real working conditions and show that their design is able to discriminate acceptance values of the two tested suits distinctly.

The consideration of ethical values has become a major issue in any technology development processes in the past twenty years. Yet ethics is sometimes treated as a mere philosophical reflection that happens either outside or after some innovation and is often considered to require a board of experts who provide ethical judgement. *Karsten Weber* expounds a unique model for ethical evaluation that can be applied during a project and does not necessarily need additional experts from outside. The author generalizes the known MEESTAR model in this chapter with the purpose to make it applicable to any technical system and the evaluation of its ethical impacts.

This part of the book is concluded with general thoughts on legal valuation of support technologies, especially robots. *Susanne Beck* summarizes legal issues revolving around responsibility. The ascription of responsibility to technical systems (also discussed in part II of this book) poses a major problem for defining legal frames when robots are to be developed and licensed but also for jurisdictional decisions that have to be made in the context of governing law. Beck examines above all the case of damages and mistakes. In conclusion, she picks up the controversial subject of transferring decision authority and thus responsibility to autonomous machines and argues for pragmatic solutions in each particular case instead of looking for overall regulations.

The subjects of this last part of the book mark also the beginning of further analyses. Some of the articles of this subsection do also provide ideas and instruments that might be used to further analyze demands and expectations, to modify designs and constructions, or to rethink the forms and contexts of deployment for support technologies. In any case, this book has a linear structure due to being a book. Yet it evolves in circles and cross-references to display the circular and networked relations of technical development.

Musculoskeletal Simulation and Evaluation of Support System Designs

Jörg Miehling, Alexander Wolf and Sandro Wartzack

Abstract Simulation of musculoskeletal models is getting more and more attention in gait analysis and surgical planning procedures. However, there are plenty other applications, where these can be used to facilitate better product and process designs. Musculoskeletal models offer the possibility to investigate and design human-technology interactions. In contrast to most conventional, empirically-based methods and tools used for workplace design, musculoskeletal models enable to catch a glimpse into the human body, revealing the inner strain conditions necessary to counteract the external loads resulting from the task to be performed. This contribution shows approaches to model human-technology interactions for support system design and directions on how to simulate, evaluate, and optimize these by means of musculoskeletal simulation as a virtual human factors tool. Finally, future prospects in musculoskeletal simulation of support devices are given.

1 Motivation

Ergonomics or human factors engineering specifically focuses on work processes and systems in conjunction with human workers. Despite ever increasing automation, there are still many manual tasks in industry, which can be accomplished efficiently only by human workers. There is a rising heterogeneity of human competencies in developed societies due to the prevailing demographic change. This effect combined with highly repetitive tasks and high loads leads to a high prevalence of work-related musculoskeletal disorders (WMSD) [Cos10], making at least some kind of

J. Miehling (✉) · A. Wolf · S. Wartzack
Engineering Design, Friedrich-Alexander-Universität Erlangen-Nürnberg, Martensstraße 9, 91058 Erlangen, Germany
e-mail: miehling@mfk.fau.de

A. Wolf
e-mail: a.wolf@mfk.fau.de

S. Wartzack
e-mail: wartzack@mfk.fau.de

© Springer Nature Switzerland AG 2018
A. Karafillidis and R. Weidner (eds.), *Developing Support Technologies*, Biosystems & Biorobotics 23, https://doi.org/10.1007/978-3-030-01836-8_21

workplace assessment a legal necessity. Various technologies are being developed to support workers to improve work results and safety at work. The development of such supportive devices, however, comprises various challenges such as the adaption of the interaction between the system and the workers [Kar15, Wei13].

Powerful virtual human factors tools to analyze the kinematics and kinetics of the human movement apparatus are musculoskeletal simulations. These tools are capable to reveal the internal reactions that occur while performing a specific task. Internal processes in question range from joint reaction forces and joint torques to neuromuscular activation in terms of inter- and intramuscular coordination as a consequence of a given motion behavior and external loads acting on the human.

We propose the application of musculoskeletal simulations to harness the abilities of such simulations in the design and evaluation of support technologies. Depending on the current design stage, such models can be used to computationally identify the level of support systems, but also be coupled with virtual prototypes of supportive devices facilitating musculoskeletal simulation to virtually evaluate support system designs. We assume that this virtual product and process development approach enables an evaluation and validation of the consequences of design choices already in early design stages leading to better and safer designs in conjunction with the reduction of physical user tests and consequently shortened design cycles [Mie13a].

2   Fundamentals of Musculoskeletal Simulation

2.1   Stress and Strain Concept

The stress and strain concept [Roh84] is a theoretical approach to describe and analyze the cause-effect relationships between human beings and its environment. This concept addresses the inner reactions of the human body to counteract the imposed external conditions satisfying Newton's laws of motion. In ergonomics physical, organizational, or social stresses can be relevant. The strain is the psychophysical or behavioural reaction of the human being as a function of the stresses itself as well as the individual's competences and capabilities. Consequently, equal stress yields diverse strain in different people. It has to be mentioned, that direct measurement of the inner strain conditions is very limited. Only few and mostly indirect indicators like heart rate, breath rate, or muscular activity can help to gain insights about the strain prevalent in a specific person [Sch10].

Most conventional methods in ergonomic product and process engineering, like REBA, RULA, the NIOSH lifting equations, or the OWAS worksheet [Dem05] are based on empirical findings and do not represent and model the human interior structure. These assessment methods are therefore restricted to the conditions prevalent in the underlying study. In contrast, musculoskeletal models enable to unveil the inner strain reactions regarding the dynamics of movement and the external boundary conditions in a holistic virtual systems engineering.

2.2 Musculoskeletal Simulation

Musculoskeletal simulation is based on multibody dynamics. A multibody system defines the human movement apparatus, with rigid bodies representing the bones or body segments. These are interconnected by mechanical joints approximating the anatomical degrees of freedom of the human body. The muscles act between the bones (segments) as unilateral actuators following the hill model [Hil38, Zaj89]. The basic musculoskeletal model commonly represents an average individual. Scaling tools allow, under consultation of empirical anthropometric data, the adaption of the models' size, weight, inertia, mobility, and strength parameters to specific users or user groups [Mie15a, Mie13b]. The underlying simulation systems are able to carry out dynamic analyses of the human musculoskeletal system under consideration of muscle and joint strain as well as neurophysiological reactions [Mie13a].

To do so, the differential equations of motion can be solved with two different computational approaches [Zaj93]. Both are applicable to solve the muscle recruitment problem, which arises from the fact, that there are more muscles than degrees of freedom in the human body. *Forward dynamic tracking algorithms* compute movement based on initial guesses of muscle activations and external forces via an integration of the equations of motion. An optimization problem distributes the forces in such way that a certain movement is computed. This approach is only applicable to certain cases and is therefore rarely used. In addition to the fact that the initial assumptions of muscle activation are hardly available or measurable, the forward dynamic approach contains a large number of unknown variables that lead to long computation times. The most common approach is *inverse dynamics*. Inverse dynamics uses known external forces and moments imposed on the human body and kinematic data (movement) as input to compute the muscle and joint reaction forces. In inverse dynamics, an optimization problem determines the distribution of muscle forces necessary to perform the given movement while producing the given external reactions. This optimization problem can be solved efficiently, making the inverse dynamics approach computationally less demanding [Cro81, Ras01].

The standard workflow of musculoskeletal simulation involving inverse dynamic simulation contains preceding measurement of the motion behaviour and the external forces. These are commonly gathered in specialized motion laboratories consisting of a motion capture system [And09] in combination with force plates [Jun14]. Inverse Kinematics is used to compute joint angle curves from the marker trajectories in order to animate the digital human model according to the measured motion. Examples of musculoskeletal modeling and simulation environments are the commercially available AnyBody Modeling System [Dam06] and the open-source tool OpenSim [Del07].

3 Virtual Development and Evaluation of Support System Designs

Musculoskeletal simulation was not primarily devised for product development, but rather for human movement science or clinical applications, for example the simulation of pathological gait in order to support surgical planning processes [Fox09]. Nevertheless, musculoskeletal models hold high potential to analyze technology that strongly interacts with humans. This is due to their ability to give insight into the biomechanical system of the human body. Supportive devices, like exoskeletons, inherently exhibit a very high level of interaction. Human and device literally merge to one integrative system performing the given tasks collaboratively.

Information gathered through musculoskeletal simulation may support the development of support devices depending on the design stage. The user-centred design process of [DIN11] helps to expediently integrate musculoskeletal simulations into the design cycle of support systems.

In the first phase, the *context of use specification*, the people performing a specific manual task and the task itself for which a supportive device is to be created have to be identified. Following the mentioned standard workflow of musculoskeletal simulation, motion capture of the unsupported task can be done while empirically assessing the workplace. The task in question can be reconstructed alternatively in a motion laboratory. In the subsequent *requirement specification phase* musculoskeletal simulation may reveal a glimpse on the forces acting in the human body while performing this task. Result of this step may be critical muscle and joint reaction forces, which are consulted to assess the need and determine the level for support. This is illustrated in the following example. A passive support system for the (orthopedically correct) lifting of heavy objects in production is to be developed. In the

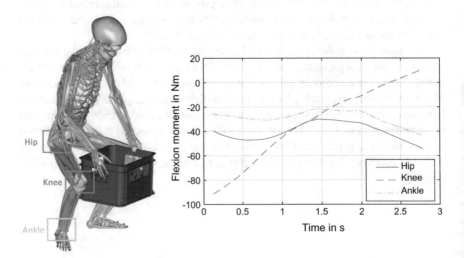

Fig. 1 Flexion moments in certain joints needed to perform the lifting task

requirement specification phase, the unsupported lifting movement is analyzed. This reveals that the muscles have to produce an initially high and linearly decreasing moment in the knee joint (Fig. 1). Therefore, the knee joint is predestined to be supported by a torsion spring.

The next phase addresses the *creation of design solutions* of the supportive device. This design stage begins with the development of rough concepts. Therefore, it is possible to add abstract forces or moments to several joints or body parts of the human model in order to evaluate where, when, and to what extent the human body should be supported [Dem17]. This information helps in the derivation of principal topological solutions for the support system fulfilling the previously specified requirements. In the example above, various abstract moments could be attached to the hip, knee, or ankle joint in order to analyze their reciprocal positive and negative effects on the movement. To keep it simple we just added a moment to the knee joint. The moment is modeled as a torsional spring defined by the torsion elastic modulus K and the twisting angle (knee flexion angle). In this way, it can be determined to what extent a certain elastic modulus K may support the movement. The support is for instance expressible in a reduction of the metabolic energy rate (Fig. 2) or the joint torque reduction relative to the unassisted task.

The concepts resulting from the previous analyses can be implemented as digital mock-ups. These virtual prototypes have to provide the kinematics and kinetics of the design concepts. To virtually *compare and evaluate* the different support system designs in conjunction with the use case in question, the digital mock-ups can be coupled with the biomechanical digital human model to simulate the interaction in detail. Concepts to integrate the digital mock-up of the support system with the digital human model are given in the next chapter. Explorative parametric studies may reveal the influence of product parameters on the inner body strains and lead to an optimized device design fulfilling the specified requirements.

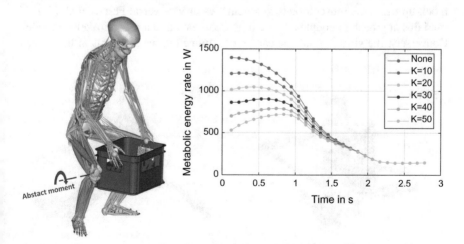

Fig. 2 Reduction of the metabolic energy rate via an added abstract moment

After the virtual development, it is recommended to build a physical prototype to assess the developed support technology in the real-world setting. This procedure assures that no unforeseen negative effects occur in consequence of the support and that the support technology meets the requirements also in the real use case. As the motion behavior is expected to change due to the intrusion of the support technology, musculoskeletal simulation should be seen as an extension to conventional biomechanical approaches rather than a surrogate.

4 Modeling of Human-Technology Interaction

A major challenge of evaluating and optimizing a support device's design using musculoskeletal simulation is the modeling of the human-technology interaction, in order to couple digital mock-ups with the musculoskeletal model. In a first step, the digital mock-up has to be modeled as multibody model including its kinematic and kinetic properties (rigid bodies in conjunction with spring elements or active actuators). Subsequently, either the musculoskeletal model needs to be imported to the CAD environment [Kru14], or the digital mock-up is to be imported into the musculoskeletal simulation environment. Of course, the digital mock-up can also be approximated in the musculoskeletal simulation environment. This way Agarwal et al. [Aga13] optimize the design of a rehabilitation finger exoskeleton. In the majority of cases however, the mock-up needs to be coupled to the digital human model. One way to achieve this is by point to point constraints and rigid connections as displayed in Fig. 3 at the example of a rower attached to a skiff [Mie15b]. This approach can be conveyed to support systems attached to virtual humans.

Support devices are often mechanisms parallel to the human body, with identical rotational axes. Therefore, it is useful to describe the kinematics of the digital mock-up using the musculoskeletal kinematics or vice versa. Ferrati et al. [Fer13] used this approach to couple a lower limb exoskeleton to a musculoskeletal model. Conversely, the virtual prototype can be constrained to the segments of the human

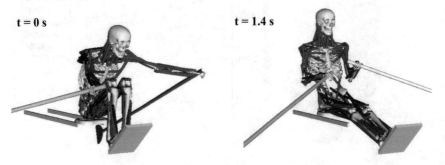

t = 0 s   t = 1.4 s

Fig. 3 Musculoskeletal model interacting with skiff through point to point constraints (seat—seat-bones) and rigid connections (footstretcher—feet; oar handles—hands)

model in such a way, that the prescribed movement of the human body defines the mock-up's kinematics. Zhou and Li [Zho16] present this approach for the design optimization of a passive upper limb exoskeleton. If the kinematic properties of a device are not expressible via the human kinematics, additional movement specification is necessary. These constraints integrate the digital mock-up of the support system with the human multibody system. The actuator forces/controls of the support system can either be predefined or optimized during the inverse dynamic simulation. The interaction forces are automatically computed as constraint forces between the human body and the supportive device in order to reach a dynamic equilibrium. Besides the use of kinematic constraints, also dynamic constraints like contacts [Kru17] can be deployed to connect human and support system models.

5 Future Prospects

This contribution shows how to design and compare support technology designs by harnessing the strength of musculoskeletal simulation. However, further research is needed in order to enhance the ease of use and efficiency of these simulation systems for product and process design. Using the conventional workflow of biomechanical simulation, still some kind of motion capture is needed. Therefore, the physical workspace and tools to interact with are necessary for such evaluations. Furthermore, motion strategies are expected to be notably different in assisted tasks in comparison to unassisted execution. Novel motion synthesis approaches like the use of motion libraries in conjunction with kinematic optimization or fully synthetic task-based optimal control strategies hold the potential to further reduce the need for physical prototypes and extensive observations prior to the musculoskeletal simulations [Mie13a, Wol17]. Of course, such simulation approaches do not replace the prototypes completely, but enable frontloading in the development process and may lead to better designs prior to physical testing. Finally, the resulting behaviour of the developed assistive technology has to be evaluated in real-world user tests to ensure the intended behaviour and safety. In addition, Young and Ferris [You17] identified the control of current exoskeletons as a major weakness. Real-time musculoskeletal simulations may reveal novel control approaches for smart active supportive devices. Lastly, since musculoskeletal models are usually rigid multibody systems, it is not possible to analyze skin pressure or soft tissue deformation that usually arise when fixing an exoskeleton to the human body. An integration or co-simulation of human finite element and musculoskeletal analysis may achieve this.

References

[Aga13]   Agarwal, P., Kuo, P.-H., Neptune, R. R., & Deshpande, A. D. (2013). A novel framework for virtual prototyping of rehabilitation exoskeletons. *IEEE International Conference on Rehabilitation Robotics.* 1–6

[And09]   Andersen, M. S., Damsgaard, M., & Rasmussen, J. (2009). Kinematic analysis of over-determinate biomechanical systems. *Computer Methods in Biomechanics and Biomedical Engineering, 12*(4), 371–384.

[Cro81]   Crowninshield, R. D., & Brand, R. A. (1981). A physiologically based criterion of muscle force prediction in locomotion. *Journal of Biomechanics, 14*(11), 793–801.

[Cos10]   da Costa, B. R., & Vieira, E. R. (2010). Risk factors for work-related musculoskeletal disorders: A systematic review of recent longitudinal studies. *American Journal of Industrial Medicine, 53*(3), 285–323.

[Dam06]   Damsgaard, M., Rasmussen, J., Christensen, S. T., Surma, E., & de Zee, M. (2006). Analysis of musculoskeletal systems in the AnyBody modeling system. *Simulation Modelling Practice and Theory, 14*(8), 1100–1111.

[Del07]   Delp, S. L., Anderson, F. C., Arnold, A. S., Loan, P., Habib, A., John, C. T., et al. (2007). OpenSim: Open-source software to create and analyze dynamic simulations of movement. *IEEE Transactions on Biomedical Engineering, 54*(11), 1940–1950.

[Dem05]   Dempsey, P. G., McGorry, R. W., & Maynard, W. S. (2005). A survey of tools and methods used by certified professional ergonomists. *Applied Ergonomics, 36*(4), 489–503.

[Dem17]   Dembia, C. L., Silder, A., Uchida, T. K., Hicks, J. L., & Delp, S. L. (2017). Simulating ideal assistive devices to reduce the metabolic cost of walking with heavy loads. *PLoS ONE, 12*(7), 1–25.

[DIN11]   German Institute for Standardization DIN EN ISO 9241. (2011). Ergonomics of human-system interaction—Part 210: Human-centred design for interactive systems. Berlin: Beuth.

[Fer13]   Ferrati, F., Bortoletto, R., & Pagello, E. (2013). Virtual modelling of a real exoskeleton constrained to a human musculoskeletal model. In N. F. Lepora, A. Mura, H. G. Krapp, & P. F. M. J. Verschure (Eds.), *Biomimetic and biohybrid systems* (pp. 96–107). Berlin: Springer.

[Fox09]   Fox, M. D., Reinbolt, J. A., Õunpuu, S., & Delp, S. L. (2009). Mechanisms of improved knee flexion after rectus femoris transfer surgery. *Journal of Biomechanics, 42*(5), 614–619.

[Hil38]   Hill, A. V. (1938). The heat of shortening and the dynamic constants of muscle. *Proceedings of the Royal Society B: Biological Sciences, 126*(843), 136–195.

[Jun14]   Jung, Y., Jung, M., Lee, K., & Koo, S. (2014). Ground reaction force estimation using an insole-type pressure mat and joint kinematics during walking. *Journal of Biomechanics, 47*(11), 2693–2699.

[Kar15]   Karafillidis, A., & Weidner, R. (2015). Grundlagen einer Theorie und Klassifikation technischer Unterstützung. In R. Weidner, T. Redlich, & J. P. Wulfsberg (Eds.), *Technische Unterstützungssysteme* (pp. 66–89). Berlin: Springer.

[Kru14]   Krüger, D., & Wartzack, S. (2014). Towards CAD integrated simulation of use under ergonomic aspects. In *Proceedings of the International Design Conference—DESIGN 2014* (pp. 2095–2104).

[Kru17]   Krüger, D. & Wartzack, S. (2017). A contact model to simulate human-artifact inter-action based on force optimization. Implementation and application to the analysis of a training machine. *Computer Methods in Biomechanics and Biomedical Engineering, 20*(15), 1589–1598.

[Mie13a]   Miehling, J., Krüger, D., & Wartzack, S. (2013). Simulation in human-centered design—Past, present and tomorrow. In M. Abramovici & R. Stark (Eds.), *Smart product engineering* (pp. 643–652). Berlin: Springer.

[Mie13b] Miehling, J., Geißler, B., & Wartzack, S. (2013). Towards biomechanical digital human modeling of elderly people for simulations in virtual product development. *Proceedings of the Human Factors and Ergonomics Society Annual Meeting, 57*(1), 813–817.

[Mie15a] Miehling, J., & Wartzack, S. (2015). Strength mapping algorithm (SMA) for biomechanical human modelling using empirical population data. In *Proceedings of the 20th International Conference on Engineering Design—ICED 15* (pp. 115–124).

[Mie15b] Miehling, J., Schuhhardt, J., Paulus-Rohmer, F., & Wartzack, S. (2015). Computer aided ergonomics through parametric biomechanical simulation. In *ASME 2015 International Design Engineering Technical Conferences and Computers and Information in Engineering Conference.*

[Ras01] Rasmussen, J., Damsgaard, M., & Voigt, M. (2001). Muscle recruitment by the min/max criterion—A comparative numerical study. *Journal of Biomechanics, 34*(3), 409–415.

[Roh84] Rohmert, W. (1984). Das Belastungs-Beanspruchungs-Konzept. *Zeitschrift für Arbeitswissenschaft, 38*(4), 193–200.

[Sch10] Schlick, C., Bruder, R., & Luczak, H. (2010). *Arbeitswissenschaft* (3rd ed.). Heidelberg: Springer.

[Wei13] Weidner, R., Kong, N., & Wulfsberg, J. P. (2013). Human hybrid robot. A new concept for supporting manual assembly tasks. *Production Engineering, 7*(6), 675–684.

[Wol17] Wolf, A., Miehling, J., & Wartzack, S. (2017). Vorgehensweise zur Vorhersage menschlicher Bewegung durch muskuloskelettale Simulation. In *Proceedings of the 28th Symposium Design for X—DfX 2017* (pp. 13–24).

[You17] Young, A. J., & Ferris, D. P. (2017). State of the art and future directions for lower limb robotic exoskeletons. *IEEE Transactions on Neural Systems and Rehabilitation Engineering, 25*(2), 171–182.

[Zaj89] Zajac, F. E. (1989). Muscle and tendon: Properties, models, scaling, and application to biomechanics and motor control. *Critical Reviews in Biomedical Engineering, 17*(4), 359–411.

[Zaj93] Zajac, F. E. (1993). Muscle coordination of movement. *A perspective. Journal of Biomechanics, 26*, 109–124.

[Zho16] Zhou, L., & Li, Y. (2016). Design optimization on passive exoskeletons through musculoskeletal model simulation. *IEEE International Conference on Robotics and Biomimetics*, 1159–1164.

Space-Game: Domestication of Humanoid Robots and AI by Generating a Cultural Space Model of Intra-action Between Human and Robot

Oliver Schürer, Christoph Müller, Christoph Hubatschke
and Benjamin Stangl

Abstract Technical perception systems exhibit essential differences in comparison with human perception systems. Technical perception systems comprise geometry, numbers, and images. But humans can define only a very small portion of space by means of technically abstract values. Far more important are topologies of personal meanings rooted in cultural meanings. This leads to several problems. Humanoid robots are endowed with complex technical perception systems, unfolding a paradox: They are being developed for the most intimate areas of human existence, but they cannot participate in the human sphere of perception. Therefore, we have developed an approach that connects robotic and human perception systems. Humanoid robots are then understood as "*companions*". The object of our approach is to develop a cultural model of space, involving robots, AI, and humans within the same context of meaning.

O. Schürer (✉) · C. Müller · B. Stangl
Department for Architecture Theory and Philosophy of Technics, Vienna University of
Technology, Vienna, Austria
e-mail: schuerer@tuwien.ac.at

C. Müller
e-mail: mueller@attp.tuwien.ac.at

B. Stangl
e-mail: benjamin@stangl.eu

C. Hubatschke
Department of Philosophy, University of Vienna, Vienna, Austria
e-mail: christoph.hubatschke@univie.ac.at

© Springer Nature Switzerland AG 2018
A. Karafillidis and R. Weidner (eds.), *Developing Support Technologies*, Biosystems &
Biorobotics 23, https://doi.org/10.1007/978-3-030-01836-8_22

1   Sociocultural Starting Position

Everyday human life is being increasingly enriched by a wide range of different technologies. As artificial intelligence is already being used in areas such as the stock market, face recognition, or for mobile telephone interfaces, robots are now also more and more in the center of international AI-research agendas. In everyday life, the fusion of such technologies is designed to offer services which satisfy individual needs and promises and solve some social problems at the same time—for example robots being sexual partners and also providing elderly care.

Our ageing society makes increased efforts to develop humanoid robots that take over care functions. Anthropomorphic machines are seen as being able to support humans as they can for example move "*naturally*" in the intimate physical and social spaces where care and assistance take place. Crucial for the acceptance of robot assistants are their social abilities and their ability to learn independently of their surroundings [Duf03, Dau07].

Technological development is facing a paradox: Robots are being developed for the most intimate parts of human existence, yet they cannot take part in human perception space.

2   Space, Perception, and Technologies

Technical services are achieved by cross-linking automation-controlled sensors with the algorithmic processing of sensor data in software data models. The interaction of these components creates technical perception systems. This kind of technology is used in applications from the simple regulation of room temperature to facilitating the autonomous behavior of hyper-complex machines such as humanoid robots. This example also makes us realize how useful the different technical perception systems are and how their different objectives generate different forms of realities.

In a simple system, only the air temperature has to be connected with the heating capacity. But in a sensor system for an artificial intelligence, "*the world*" has to be mapped. The complexity and unpredictability of the everyday world in which humans usually move about effortlessly represents a huge challenge for hyper-complex technical systems. But artificial intelligence must content itself with the barriers posed by its technical perception systems and data models. Data can only be acquired and processed according to a model in line with the given perception systems. The characteristics of services the technological system can provide are a product of these circumstances. Thus, technical systems exist and operate in other realities than humans. The numerous technical perception systems differ essentially from those of humans. They perceive space in the form of geometries, numbers, and images. Humans, on the other hand, determine space only to a very small degree by means of technical values. Substantially they perceive space through personal experiences, association, and habit, each rooted in cultural meanings. Although robots possess the most modern perception systems, they do not share the same perception space as humans.

The different state-of-the-art approaches have in common that they solve problems by focusing on given technical possibilities. However, this leads to the fact that the cultural importance of objects is taking a back seat compared to the data-driven technical representation of objects. Individual meanings disappear as a result.

This is where the problems culminate that our approach tries to solve.

3   A Research Method for the Human Lifeworld

We are working on developing a cultural concept of space for humanoid perception systems. To this end, we use the abilities of a humanoid robot for human-like inter-action as an approach to data collection and modeling. In the space model presented here for discussion, the meaning of objects is inquired through language interaction and interlinked by using algorithms of machine learning. By collecting and inter-linking many localized meanings, a space model is created that cannot be reduced to technical parameters or human perceptions or individual meanings. Instead, a hybrid space model is created that is generated jointly by means of natural language inter-action. This is done in order to facilitate new interaction spaces equally developed by humans and robots. Thus, a humanoid robot with its artificial intelligence system is involved in the same meaning context as we human beings are.

The term *language-game* has been coined by Ludwig Wittgenstein [Wit01]. It means that every linguistic expression is rooted in human life. Only there do the differ-ent human language-games make sense. Every word, every term, and every sentence have meanings which depend on the context or actions or situations in which they are expressed. Mathematics and formal logics, too, are included in language-games. Just as the philosophical language-game, our architectural space-game also uses the con-nection between linguistic expressions and human practices. By contrast, however, we are dealing with the cultural relationship between humans and humanoids to con-stitute everyday space. Using naturally linguistic man/machine communication, we want to facilitate the mutual construction of space and meaning in dialogue. Within the possibilities of humanoid shape, we reproduce human gestures and body stances to enrich this construction with non-linguistic communication. With the help of such interaction, perception space is created for humanoids which is formed interactively from meanings, far beyond any technical parameters.

The human lifeworld is essentially spatial. Space in this sense is mainly a holistic substrate of meanings and their localization. All relations discussed in social and cul-tural space theories arise from these localized meanings. Architecture is specifically dedicated to aspects of space in connection with life. It faces changes which will fundamentally transform behaviors (regarding the production and use of spaces) that have long been culturally conserved. Buildings as we know them will be increasingly equipped with mobile parts and sophisticated controls. We are on the threshold of a development where technological artifacts such as the home and the robot will interact and form everyday perception spaces.

The effective forces of each reality are competing for the prerogative of interpretation regarding the actuality of the everyday world. This structural otherness of human and technological realities harbors the risk of conflict and danger. Often users of assistance systems feel harassed, pushed, or under-challenged and reject the otherness for that reason.

Man's social spaces are manifold. Social space is not a permanently fixed space but is being constantly *produced* [Lef91] and is therefore constantly changing. The production of space is never a neutral process but always co-determined by power structures, economic interests, and cultural hegemonies. Like space, technologies should also not be understood as being neutral because they are also involved in power structures. Our project is trying to address these questions of power and to promote the authority of users with regard to artificial intelligence. At the same time, we are addressing *technical objects* [Sim12] as active actors. Humanoid robots are not simply other objects or neutral actors, but must be theorized and investigated as active co-designers of such spaces. Furthermore, it must always be taken into account in what kinds of power structures these humanoids are being developed, researched, and utilized. Just like Langdon Winner did in his essay *Do Artifacts have Politics?* [Win87] we also want to inquire what cultural, ethical, and political concepts are built into such robots, especially in relation to architectural spaces that form our domestic environment.

These questions are to be processed and philosophically reflected, using transdisciplinary methods and artistic research. New transdisciplinary kinds of access and methods are to be developed and involved in current debates and discourses about questions of robot ethics [Lin15]. Artistic research allows us—going beyond technical questions and questions of social science—to develop, evaluate, and reflect upon the question of the value of cultural meanings of space, the different kinds of space perception, and possibilities of a common understanding of space for robots and non-robots. The findings and insights will not be evaluated in relation to social or technical objectives in their cause-and-effect connection, but as an open result in lifeworld contexts.

How spaces are designed, changed, and produced, and how we move in these spaces and interact with each other, is influenced largely by our perception. What we see, hear, feel, etc. is co-determined by our lifeworld. However, how objects, spaces, and people are seen, heard, and felt is governed to a considerable degree by cultural factors. The cultural meanings of spaces and objects are permanently negotiated and transformed. Humanoids are co-producers of social space and they interact with space as much as with humans in space. Therefore, humanoid robots should not only learn to recognize these cultural meanings, but also to co-develop them. Doing that, they will occupy a novel position between existing technology and humans.

4 Relationship of Humans and Technological Artefacts

What principles are useful for developing the relationship between unique human individuals and humanoids—a serially produced, distorted, technical wanna-be mirror image of humans?

It was a relatively simple and everyday situation from which the French philosopher Jacques Derrida developed complex considerations and profound reflections. One day, just as Derrida stepped out of the shower, he realized that he stood naked in front of his cat which had sneaked into the bathroom. What fascinated Derrida in this encounter was the simple observation that he was embarrassed to be exposed to the cat's look in this situation. Although the cat surely had no idea what nakedness means and most likely showed no interest in the fact that he was naked, Derrida felt observed by the cat's glance. "How can an animal look you in the face?" Derrida asked in his talk called *The Animal That Therefore I Am*, in which he reflected this situation [Der02].

In the same way as Derrida's cat, humanoid robots for domestic service are constructed to look us in the face—to be able to see us. It almost seems as if this ability has been one of the main reasons for developing humanoid robots at all: that robots have a face which we can look at, but also a face that can look back at us, that is able to return our look [Coe11, Coe12, Coe14]. With cameras as eyes, humanoid robots are not only able to "see", but also to record. In addition, they are equipped with a large number of other sensors which continuously track and evaluate their surroundings. They are also connected with numerous other helping and assisting technical artifacts. All this produces a completely different perception of space, time, and actions than humans have who are socioculturally conditioned and architecturally influenced. Should we feel shame at a time when humanoid robots are advancing more and more into our private and most intimate spaces, when they are able to see us and look at us? Or are these robots only a collection of technically-based offers in a system they represent, mere technical servants whose glance does not touch us at all? Or are they actually a kind of companion—hybrid beings that share the most intimate spaces with us while potentially they could also share these worldwide on the Internet?

In her book *When Species Meet* [Har08] the American philosopher and theoretician Donna Haraway criticizes Derrida's approach. She argues that he has failed to recognize the many possibilities and meanings of the cat's glance. According to Haraway, this look of the cat is an invitation to a joint becoming, a *becoming with* what she calls a "companion species". Every look is a reciprocal process and living with another being is also a common becoming. According to Haraway, this common becoming also requires another kind of access to ethics, namely an ethics of shared responsibility. Just as a guide dog is responsible for its master or mistress, the latter are also responsible for the dog. Living with a dog is based on a mutually shared responsibility for each other. The following principle applies not only to the man/dog relationship: "Responsibility is a relationship crafted in intra-action through which entities, subjects and objects, come into being" [Har08, p. 71]. Intra-action is understood as the mutual constitution of entangled agents [Bar07].

Can technical artifacts in a relationship with humans be regarded as the kind of "companion species" that Haraway defines? What could such a mutually shared responsibility between human and robot look like? Based on a similar understanding of mutually shared responsibility, we want to reflect on the ethical, political, and spatial consequences and develop a technical system in which humans care for their robots who in turn support these humans in everyday life.

5 Towards a Human-Humanoid Lifeworld

The state of current research is that technical perception systems translate objects and users into numbers, register spaces in geometric terms and display them as images. However, they are unable to place these elements in their environment into a mutual relationship that is meaningful for humans. By contrast, we have developed an approach in which the design of a model for the cultural meaning of space is based on interaction between humans and robots. The machine is regarded as a *companion*. This approach allows the technical system to use the interactive requisition of cultural meanings to develop a concept of space and to adapt it dynamically—thus domesticating humanoid robots and AI.

To research humanoid robots as a potential future *companion species* also means to integrate the above questions according to new forms of perception. The allocation of meanings to objects, spaces, building components, persons, and machines is a network-like reference system that is woven from geometrical, pictorial, social, and cultural relationships. Meanings of the individual parts and their relationships to each other are constantly and dynamically transformed through negotiations. Humonoids are equipped with sensors that produce completely different data of the environment than the human perception apparatus. Transdisciplinary, artistic research can help to make these specific forms of perception visible, and also to show possibilities of interweaving the intra-action between humanoid and human kinds of perception in order to generate a cultural space model of life: A model of space that can neither be reduced to the mere human perspective nor to the mere humanoid perspective.

References

[Bar07]   Barad, K. (2007). *Meeting the universe halfway: Quantum physics and the entanglement of matter and meaning*. Durham: Duke University Press.
[Coe11]   Coeckelbergh, M. (2011). Humans, animals, and robots: A phenomenological approach to human-robot relations. *International Journal of Social Robotics, 3,* 197–204.
[Coe12]   Coeckelbergh, M. (2012). Growing moral relations: Critique of moral status ascription. Palgrave Macmillan.

[Coe14] Coeckelbergh, M., & Gunkel, D. J. (2014). Facing animals: A relational, other-oriented approach to moral standing. *Journal of Agricultural and Environmental Ethics, 27*.

[Dau07] Dautenhahn, K. (2007). Methodology and themes of human-robot interaction: A growing research field. *International Journal of Advanced Robotic Systems*.

[Der02] Derrida, J. (2002). The animal that therefore I am (More to follow). *Critical Inquiry, 28*(2), 369–418.

[Duf03] Duffy, B. R. (2003). Anthropomorphism and the social robot. *Robotics and Autonomous Systems, 42*(3), 177–190.

[Har08] Haraway, D. (2008). When species meet. University of Minnesota Press.

[Lef91] Lefebvre, H. (1991). The production of space. Blackwell.

[Lin15] Lin, P., Abney, K., & Bekey, G. A. (Eds.). (2015). Robot Ethics. MIT-Press.

[Sim12] Simondon, G. (2012). Die Existenzweise technischer Objekte. Zürich, diaphanes.

[Win87] Winner, L. (1987). *Do artifacts have politics?. The whale and the reactor: A search for limits in an age of high technology*. Chicago: University of Chicago Press.

[Wit01] Wittgenstein, L. (2001). Philosophische Untersuchungen. *Wissenschaftliche Buchgesellschaft Frankfurt*.

ROS-Based Robot Simulation in Human-Robot Collaboration

Paul Glogowski, Kai Lemmerz, Alfred Hypki and Bernd Kuhlenkötter

Abstract The idea of human-robot collaboration (HRC) in assembly follows the aim of wisely combining the special capabilities of human workers and of robots in order to increase productivity in flexible assembly processes and to reduce the physical strain on human workers. The high degree of cooperation goes along with the fact that the effort to introduce an HRC workstation is fairly high and HRC has hardly been implemented in current productions so far. A major reason for this is a lack of planning and simulation software for the HRC. Therefore, this paper introduces an approach of how to implement such a software on the basis of the Robot Operating System (ROS) framework in order to enable a realistic simulation of the direct cooperation between human workers and robots.

1 Introduction

In human-robot collaboration (HRC), human workers and robots work simultaneously in different production steps in *one* common and overlapping workspace. This offers new possibilities for work planning, especially in assembly, because tasks can be processed by human workers and robots individually. By taking over non-ergonomic tasks, the robot can relieve the human worker's workloads and make workplaces more attractive. The major objective of HRC is to increase the profitability of assembly workstations. However, HRC has hardly been introduced into companies' routines. This is due to the uncertainty in the implementation of the prescribed standards. Furthermore, the economic efficiency is still unknown because many interruptions and safety requirements, such as speed reductions or safety stops, may limit the previously calculated productivity. Moreover, there is still a lack of planning support, e.g., in the form of assistance, consulting services, and especially software tools [Ben16, Glo17].

P. Glogowski (✉) · K. Lemmerz · A. Hypki · B. Kuhlenkötter
Chair of Production Systems, Ruhr-University of Bochum, Universitätsstr. 150, 44801 Bochum, Germany
e-mail: glogowski@lps.rub.de

© Springer Nature Switzerland AG 2018
A. Karafillidis and R. Weidner (eds.), *Developing Support Technologies*, Biosystems & Biorobotics 23, https://doi.org/10.1007/978-3-030-01836-8_23

A comprehensive simulation software is available for manual work, which supports the complex process of a detailed planning and implementation. In addition, there is a large number of digital human models [Tsa16] that can be integrated in the context of specific software developments or enhancements [Bus13]. With the help of these digital human models and corresponding methods from the field of ergonomics, it is possible to simulate ergonomic aspects or cycle time analyses for manual assembly tasks [Bus15, Fri10].

However, appropriate software is not yet fully available for the simulation of HRC. To meet this demand, this paper will introduce a way to implement the sub-area of robot simulation in HRC. For this purpose, *Open Source Software* (OSS) is exclusively used as there are many good and useful solutions in robotics. In addition, OSS allows a cost-efficient and manufacturer-independent approach, which makes the software future-proof and reusable because it makes use of open standards. It is based on the well-known framework *Robot Operating System*[1] (ROS), which enjoys great popularity in all parts of robotics. One of the reasons for this is that its structure and abstraction level make it suitable for a very wide range of robots.

The paper is structured as follows: Chap. 3 provides an overview of the state of the art in HRC. Chapter 4 introduces the basics of the framework ROS and gives an insight into the most important components and functionalities of ROS. Against this background, Chap. 5 demonstrates how a simplified HRC workstation with the focus on robot simulation is implemented and simulated based on ROS.

2 Human-Robot Collaboration

The following chapter outlines the existing collaborative robot hardware, the associated safety and sensor technology as well as the necessary robot simulation software tools for HRC.

Hardware

Suitable hardware is a prerequisite for the successful implementation of an HRC application. There is a number of light-weight robots (LWR) approved for HRC operations whose weight, dimensions, and load-bearing capacity are close to the capabilities and physical constitution of a human arm. The development of HRC in the field of robot hardware has made an important progress in many ways. Thus, the robot arms are equipped with more and more sensors and are therefore basically suitable to collaborate with human workers in a safe way.

Safety and Sensor Technology

In order to enable the cooperation of human workers and robots in one workspace, various safety requirements have to be met. This serves primarily to protect the

---

[1] http://www.ros.org/.

human worker who is performing the collaboration with the robot. The risk of injury from collisions of the robot arm with the human body must be minimized as much as possible. The maximum permissible forces and pressures which may occur in the event of a collision are specified, for example, in the standard DIN ISO/TS 15066 [DIN16].

An HRC application is highly dynamic: Human workers are continuously moving, workpieces can be in varying positions or a planned trajectory is blocked by an obstacle. In order to meet the high safety requirements, the collaborative workspace must be constantly monitored with appropriate sensors. Optical sensors, laser scanners, or ultrasonic sensors are available for this purpose, for example the safe camera system PILZ SafetyEYE. The position of the human worker must be known in order to adjust the speed of the robot or to stop a movement when entering different safety zones. Since the human worker and robot operate simultaneously on a workpiece in a shared workspace, the robot must permanently redefine the position of the workpieces and, if necessary, dynamically reschedule a movement. In addition to preventive safety techniques, sensors are also used to reduce consequences of collision. For example, torque or tactile sensors provide the robot with appropriate sensitivity.

Simulation Tools

In order for hardware and sensors to work together profitably, a suitable simulation tool is required. The high dynamics of an HRC application require a large number of modern algorithms which will enormously increase the software complexity. In order to guarantee a high quality of the simulation system, it is necessary to simulate the HRC workstation as close to reality as possible. The danger of testing new software solutions directly in real applications is huge in the field of HRC, especially because of the direct contact to human workers. The development of a comprehensive simulation would solve this problem by simulating not only human workers and robots but also sensors and the environment. The aim is to design the software as universally as possible for all collaborative robots, end effectors and sensors in order to generate a high benefit for many potential users.

Up to now, robot technology has been characterised by proprietary software systems that are used to simulate and control robot movement. The different robot simulation systems can be specified according to manufacturer-specific and manufacturer-independent systems. Leading robot manufacturers such as ABB and KUKA rely on proprietary software solutions and provide their own simulation systems (ABB RobotStudio, KUKA.Sim). These systems only support the control of proprietary robots. In contrast, manufacturer-independent simulation systems such as DELMIA, EasyROB and RobCAD have the advantage of providing a universality for controlling generic robot types from different manufacturers [Web17, Hyp08].

A development of the above-mentioned proprietary software is difficult or even impossible: A solution can only be implemented by the manufacturers themselves, any external development is absolutely bound to the provided software interfaces. In this context, this paper considers a solution approach for robot simulation with OSS. It is built on existing OSS since the complexity of an HRC system would otherwise be difficult to handle.

3 Robot Operating System

The most widely used open source software framework for robots is the *Robot Operating System* (ROS) [Qui09]. Today, ROS is considered as standard in robot research and development, and is universally used in robotics [Jos15a, Guz16]. ROS has a huge collection of drivers that enables the control of sensors and actuators as well as complete, commercially available robot systems. These include a wide range of basic and innovative robotics algorithms for e.g., motion planning, sensor data processing, collision checking, navigation, and many others. In addition, there is a wealth of utilities available to facilitate the visualization and simulation of a robot [Qui15, Jos15b].

3.1 Motion Planning Framework MoveIt!

The metapackage *MoveIt!* combines the latest algorithms for the mobile manipulation of robots. It offers, among other things, possibilities for motion planning, manipulation, 3D perception, kinematics, control, and navigation. The range of robots on which *MoveIt!* has already been used successfully, ranges from diverse jointed-arm robots to mobile manipulation systems and humanoid robots.

Robot Modeling, Kinematics, and Dynamics

To describe robotic models in ROS, there is the *Unified Robot Description Format* (URDF), which is based on the XML format. It helps define the kinematics and dynamics of a robot as well as determine the visual representation and the collision model. The workspace and any sensors can be specified within the URDF file [Jos15b]. Problems of direct and inverse kinematics can also be solved. The default is the solution algorithm of the *Kinematics and Dynamics Library*, which is based on a numerical Jacobi method. However, the numerical solution of the inverse kinematics can take a long time or cannot be solved. An analytical method which, for example, provides the package IKFast from the *Open Robotics Automation Virtual Environment* (OpenRAVE) [Dia10] can help here.

Collision Check

The *Flexible Collision Library* [Pan12], which is used as standard, offers the possibility to perform continuous collision checks, taking current sensor data into account.

Motion Planning

Motion planning is provided as a plugin solution. By default, *MoveIt!* is based on the *Open Motion Planning Library* (OMPL) [Suc12a]. This high-quality library includes a wide range of motion planning algorithms [Ros14] such as *Probabilistic*

Roadmap Method [Kav96], *Rapidly-expanding Random Trees* [LaV01], and *Kinody-namic Planning by Interior-Exterior Cell Exploration* [Suc12b]. The OMPL contains only abstract planners which are configured by *MoveIt!* for the respective application. *MoveIt!* automatically selects a motion planning algorithm contained in OMPL unless a specific algorithm is explicitly selected. In *MoveIt!*, the *Trajectory Processing Routines* take care of the temporal components of motion planning. Taking into account the speed and acceleration limits of individual joints, the routine calculates a suitable time-parameterized trajectory.

Control Strategies

For the robot simulation, as in reality, control strategies are necessary for the individual joints. Without these strategies, the simulated robot arm would collapse due to the force of gravity and could not be controlled. Various controllers are available in ROS, including force and torque, position, joint state, or speed control. User specific control algorithms can also be implemented [Qui09].

Planning Scene

The environment the robot is located in is stored in the so-called *Planning Scene*. This includes the state of the robot itself. The *Planning Scene Monitor* manages all the information needed to create the world in which the robot is located. It receives information about the current robot status (e.g., joint angles), sensor information, and geometry information on the environment [Chi16].

3D Perception

The *Occupancy Map Monitor* administrates the perception and creates an occupancy map of the environment using the *OctoMap* framework [Hor13]. This map is created by the use of distance sensors such as a TOF camera. The data is then filtered by the *Depth Image Occupancy Map Updater* so that the robot parts in the camera's field of view are not taken into account. At the end, a map is built up consisting of many small blocks representing its environment. It is considered by particular fields such as motion planning and thus helps to avoid collisions with units in its environment.

3.2 Robot Simulator Gazebo and Visualization with RViz

Gazebo is an open source 3D simulation software for robot simulation. Since ROS is only a platform for the development of robot software, but does not contain a simulator, it is necessary to use additional software like Gazebo [Koe04]. As a decoupled simulator, Gazebo has no integrated component for controlling a robot and is therefore dependent on external software such as ROS.

In Gazebo, there are four different physics engines available that allow a realistic dynamic simulation. These include the *Open Dynamics Engine*, the *Bullet Physics*

Library, *Simbody*, and the *Dynamic Animation and Robotics Toolkit*. The physics engines calculate, among other things, body collisions and occurring forces of rigid bodies. The Bullet Physics Library is also able to simulate deformable bodies.

Gazebo is able to generate realistic sensor data. It can, among others, simulate laser scanners, 2D and 3D cameras, TOF cameras, contact sensors, and force-torque sensors. There is also the option to imitate sensor noise.

The GUI application *ROS Visualization* (RViz) is one of the most commonly used tools in ROS. It offers extensive display and interaction options for a robot environment. RViz should not be compared to a simulation environment like Gazebo as it does not offer any physical considerations. Above all, RViz offers the possibility to visualize information such as sensor data. It can, for example, display the data of a TOF camera.

3.3   Applications in Industry and Research

There is already a number of projects that promise great development leaps in HRC simulation through the use of ROS. Particular mention should be made of the *LIAA* project with the participation of the Fraunhofer IPA. The aim of LIAA was to develop a framework based on ROS Industrial, which enables the cooperation of a human worker and a robot in assembly tasks. The *ReApp* project, one of LIAA's predecessor projects, shows how a profitable combination of AutomationML and ROS can be implemented [Hua16]. This is primarily intended to create benefits for SMEs through faster prototype development and reusability. In addition, Kallweit et al. [Kal16] take care of the implementation of safety requirements for collaborative robots, developing an approach based on state machines in ROS. Gradil and Ferreira [Gra16] develop a ROS-based simulation framework for human-robot interaction, in which human workers can perform simple movements. In the future, the authors want to integrate a fully moveable human model for Gazebo.

4   HRC Application

In the following chapter, an exemplary HRC workstation is simulated with the LWR UR5 and the TOF camera Xtion PRO LIVE applying Gazebo and ROS (Fig. 1).

For a well performing robot application, it is necessary to have an adequate description of the components robot (LWR UR5), sensor (TOF camera), and end effector (gripper). The TOF camera is mounted two meters high and tilted slightly forward. A simple gripper has been implemented here, since the complexity of the available gripper packages would not be appropriate at this point.

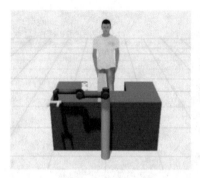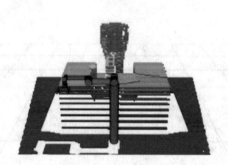

Fig. 1 Exemplary HRC application (*left*: complete environment in Gazebo, *right*: surrounding objects detected by simulated TOF camera in RViz)

4.1 Motion Simulation

In the following HRC application, the robot's TCP moves from a start to a target pose. At the target position there is an object (a can) which is to be gripped by the robot. In between there is an obstacle that is not permanently connected to the workstation. Figure 1 (left) shows the simulation environment of the HRC application in Gazebo. In the following, two different case studies are examined: one with and one without a sensor.

Motion with Switched-Off Sensors

In the first case, the TOF camera is turned off. The planning scene does not have any information about the environment, which is why it does not include any objects. Accordingly, the motion planning of MoveIt! has no knowledge of objects within its environment. Therefore, the robot's trajectory—without considering the obstacle—follows the shortest path from start to finish. In Gazebo, this causes the robot to collide with the obstacle and the object.

Motion with Switched-On Sensors

In the second case, the sensor is switched on. Figure 1 (right) shows the surrounding objects that the camera detects, presented as colored blocks in RViz. Depending on the distance of the registered points from the camera, their color changes. The environment is also integrated in the planning scene and is taken into account in the movement planning. Thus, there is no collision with the obstacle, but the path of the robot arm runs over the obstacle to the target. Consequently, collisions can basically be avoided if the TOF camera is switched on. The effect of the camera position can easily be recognized: the upper part of the human body, especially the head, is not detected, thus it would not be included in the collision checks.

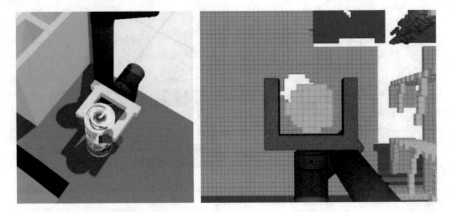

Fig. 2 Gripping process (*left*: contact points of the gripped object in Gazebo, *right*: OctoMap resolution of 15 mm in RViz)

4.2 Gripper Simulation

As a last step, this paper examines the gripping process of the object. MoveIt! offers interfaces for pick-and-place implementations and can also select the best solution from different gripper positions [Jos15b]. In order to simplify the calculation of the physical effects, the collision model of the object, which is the can in our case, is assumed to be a cuboid. Gazebo displays the contact points of the objects. Figure 2 (left) shows four blue spheres at the corners of the object's collision body with contact to the gripper. The identification and evaluation of such contact points is of particular relevance to analyze possible collisions or bruising in collaborative assembly processes. The physics engines applied in Gazebo therefore offer a promising prerequisite for the simulation-based verification and validation against the maximum permissible forces and pressures that are specified in the current standards.

5 Summary and Outlook

The application of the ROS framework has led to a planning and simulation system that combines sensor information and the motion planning of a robotic arm. A simulation environment with basic HRC functions is available. With a development based on OSS, it has been possible to implement an extensible and open application that makes the complexity of an HRC system manageable. An application of the system's current state would, of course, not be sufficient to meet the safety requirements in real operations. The aim of this paper has rather been to clarify the purpose of a software development framework in the field of robotics.

In future work, the objective will be to integrate a digital human model into the HRC application described in this paper. The focus will then be on the human movement behavior in order to plan the corresponding robot path without collisions and to determine the optimum balance of the collaborative assembly process (e.g., optimization of delivery positions). In addition, the simulative derivation of occurring collision forces in impact and bruising situations between the robot system and the human worker will be further developed in order to guarantee a safe HRC application. Finally, the results of the simulation system for HRC will be validated within the scope of real HRC applications using various demonstrators.

Acknowledgements The research and development project "KoMPI" (http://kompi.org/) is funded by the German Federal Ministry of Education and Research (BMBF) within the Framework Concept "Research for Tomorrow's Production" (fund number 02P15A060) and managed by the Project Management Agency Forschungszentrum Karlsruhe, Production and Manufacturing Technologies Division (PTKA-PFT).

References

[Ben16] Bender, M., Braun, M., Rally, P., & Scholtz, O. (2016). Leichtbauroboter in der manuellen Montage - Einfach einfach Anfangen: Erste Erfahrungen von Anwenderunternehmen.

[Bus13] Busch, F., Wischniewski, S., & Deuse, J. (2013). Application of a character animation SDK to design ergonomic human-robot-collaboration. In *Proceedings of the 2nd International Symposium on Digital Human Modeling (DHM)* (pp. 1–7).

[Bus15] Busch, F. (2015). *Ein Konzept zur Abbildung des Menschen in der Offline-Programmierung und Simulation von Mensch-Roboter-Kollaborationen* (Ph.D. thesis). University Dortmund.

[Chi16] Chitta, S. (2016). MoveIt!: An introduction. In A. Koubaa (Ed.), *Robot operating system (ROS): The complete reference* (Vol. 1, pp. 3-27). Cham: Springer International Publishing.

[Dia10] Diankov, R. (2010). *Automated construction of robotic manipulation programs* (Ph.D. thesis). Carnegie Mellon University, Robotics Institute.

[DIN16] DIN ISO/TS 15066. (2016). Roboter und Robotikgeräte - Kollaborierende Roboter.

[Fri10] Fritzsche, L. (2010). *Work group diversity and digital ergonomic assessment as new approaches for compensating the aging workforce in automotive production* (Ph.D. thesis). University Dresden.

[Glo17] Glogowski, P., Lemmerz, K., Schulte, L., Barthelmey, A., Hypki, A., Kuhlenkötter, B., et al. (2017). Task-based Simulation Tool for Human-Robot Collaboration within Assembly Systems. In T. Schüppstuhl, J. Franke, & K. Tracht. (Eds.), *Tagungsband des 2. Kongresses Montage Handhabung Industrieroboter* (pp. 155–163). Berlin: Springer.

[Gra16] Gradil, A., & Ferreira, J. F. (2016). A visualisation and simulation framework for local and remote HRI experimentation. In: *IEEE 23 Encontro Português der Computação o Gráfica e Interação (EGCGI)*.

[Guz16] Guzman, R., Navarro, R., Beneto, M., & Carbonell, D. (2016). *Robotnik—professional service robotics applications with ROS. In A. Koubaa (Eds.), Robot operating system (ROS): The complete reference* (Vol. 1, pp. 253–288). Cham: Springer International Publishing.

[Hor13] Hornung, A., Wurm, K. M., Bennewitz, M., Stachniss, C., & Burgard, W. (2013). OctoMap: An efficient probabilistic 3D mapping framework based on octrees. In *Autonomous robots* (pp. 189–206).

[Hua16]   Hua, Y., Zander, S., Bordignon, M., & Hein, B. (2016). From AutomationML to ROS: A model-driven approach for software engineering of industrial robotics using ontological reasoning. In *IEEE 21st International Conference on Emerging Technologies and Factory Automation (ETFA)* (pp. 1–8, 2016).

[Hyp08]   Hypki, A. (2008). *Beitrag zur Simulation industrieller Automatisierungssysteme* (Ph.D. thesis). University Dortmund.

[Jos15a] Joseph, L. (2015). *Learning robotics using Python* (Vol. 1). Packt Publishing.

[Jos15b] Joseph, L. (2015). Mastering ROS for robotics programming (Vol. 1). Packt Publishing.

[Kal16]   Kallweit, S., Walenta, R., & Gottschalk, M. (2016). ROS based safety concept for collaborative robots in industrial applications. In T. Borangiu (Ed.), *Advances in robot design and intelligent control: Proceedings of the 24th international conference on robotics in Alpe-Adria-Danube Region (RAAD)* (pp. 27–35). Cham: Springer International Publishing.

[Kav96]   Kavraki, L. E., Svestka, P., Latombe, J.-C., & Overmars, M. H. (1996). Probabilistic roadmaps for path planning in high-dimensional configuration spaces. *IEEE Transactions on Robotics and Automation*, 566–580.

[Koe04]   Koenig, N., & Howard, A. (2004). Design and use paradigms for Gazebo, an open-source multi-robot simulator. In *International Conference on Intelligent Robots and Systems* (pp. 2149–2154). Sendai, Japan.

[LaV01]   LaValle, S. M., & Kuffner, J. J. (2001). Randomized kinodynamic planning. *The International Journal of Robotics Research*, 378–400.

[Qui09]   Quigley, M., Conley, K., Gerkey, B. P., Faust, J., Foote, T., Leibs, J., et al. (Eds.). (2009) *ROS: An open-source robot operating system*. In ICRA Workshop on Open Source Software

[Qui15]   Quigley, M., Gerkey, B., & Smart, W. D. (2015). *Programming robots with ROS* (Vol. 1). O'Reilly Media.

[Pan12]   Pan, J., Chitta, S., & Manocha, D. (2012). FCL: A general purpose library for collision and proximity queries. *IEEE International Conference on Robotics and Automation*. (pp. 3859–3866).

[Ros14]   Rosell, J., Pérez, A.., Aliakbar, A.., Palomo, L., & García, N. (2014). The Kautham Project: A teaching and research tool for robot motion planning. In *Proceedings of the IEEE Emerging Technology and Factory Automation (ETFA)* (pp. 1–8).

[Suc12a] Şucan, I. A., Moll, M., & Kavraki, L. E. (2012). The open motion planning library. *IEEE Robotics & Automation Magazine*, 72–82.

[Suc12b] Şucan, I. A., & Kavraki, L. E. (2012). A sampling-based tree planner for systems with complex dynamics. *IEEE Transactions on Robotics* 116–131.

[Tsa16]   Tsarouchi, P., Makris, S., & Chryssolouris, G. (2016). Human-robot interaction review and challenges on task planning and programming. *International Journal of Computer Integrated Manufacturing*, 916–931.

[Web17]   Weber, W. (2017). *Industrieroboter* (Vol. 3). KG: Carl Hanser Verlag GmbH & Co.

User Acceptance Evaluation of Wearable Aids

Christina M. Hein and Tim C. Lueth

Abstract Wearable aids like exoskeletons strive to help during rehabilitation, physically demanding work in industry or nursing care, and in the military field. The evaluation of their user acceptance is crucial in order to guide the development and research and to compare different products. This chapter describes the classification of evaluations using the criteria target, type, test environment, and measuring tool. It therefore helps to classify existing studies and shows developers of evaluations different test design possibilities. Finally, a field study in nursing care is presented. Its target was to compare two different passive force assisting suits. Within this example all steps concerning the test design and analysis are shown.

1 Introduction

Wearable aids like exoskeletons are being developed for rehabilitation or physically demanding work. In order to guide their development and to compare different products, evaluation is crucial. An essential factor for their widely spread application is the user's acceptance of the wearable aid.

The term *user acceptance* describes the users' assent to the application of a product. User Acceptance Testing (UAT) is a term in the field of software testing and stands for a formal testing with respect to user needs, requirements, and business processes [Ham13]. This chapter will focus on the classification of evaluations of user acceptance for wearable aids.

C. M. Hein (✉) · T. C. Lueth
Micro Technology and Medical Device Technology, Technical University of Munich,
Boltzmannstr. 15, 85748 Garching, Germany
e-mail: christina.hein@tum.de

T. C. Lueth
e-mail: tim.lueth@tum.de

© Springer Nature Switzerland AG 2018
A. Karafillidis and R. Weidner (eds.), *Developing Support Technologies*, Biosystems & Biorobotics 23, https://doi.org/10.1007/978-3-030-01836-8_24

Wearable technology usually refers to electronic devices that can be worn on the body. This definition includes devices like hearing aids or pace makers. Though, this chapter focuses on wearable aids that show an intensive physical human-machine interface, like force assisting devices. They can be both, active driven exoskeletons or passive force assisting suits.

2   Classification of Evaluations

The evaluation has to ensure that the main aspects concerning the user acceptance of the wearable aid are assessed. This may differ, as the target group and the system itself varies. Aspects of user acceptance for specific technology applications can be found in literature. In [McC05] the acceptability of assistive technology for elderly people is described. The authors claim that there has to be a felt need for assistance, the technology has to be affordable and available, and the assistive technology has to fulfill some attributes, namely efficiency, reliability, simplicity, safety, and aesthetics.

In information systems theory several models for the acceptability exist. The technology acceptance model states that the perceived usefulness and the perceived ease-of-use influence the user to accept the technology [Dav89]. The unified theory of acceptance and use of technology was developed through the review of eight models and proposes four factors affecting the individual's acceptance: the performance expectancy (ease of use), the effort expectancy, the social influence, and the facilitating conditions [Ven91].

The key influencing factors concerning the user acceptance affect the suitable test type, test environment and measuring tools. Using these aspects, a classification of evaluations is done within this contribution. Table 1 shows an overview of classification criteria of evaluations, which are described in more detail below.

2.1   *Target of the Evaluation*

A variety of reasons may lead to the need for an evaluation. A manufacturer or developer of a wearable aid may target to identify product characteristics or decide to compare a new product with the state of the art in order to identify advantages and disadvantages of the product and derive areas for improvement. Researchers can use the comparison with the state of the art to show the novelty of the innovation.

Users usually compare different aids, often various brands, in order to identify products fitting to their specific application and environment. A systematic process to identify requirements for industrial production can be found in [Goe16].

Table 1 Classification criteria: the main aspects for user acceptance influence the test design. Tests can be grouped by their target, design type, test environment and measuring tool.

Main aspects for user acceptance	
Target	– Identify product characteristics – Compare new product with state of the art – Compare different products

Type [Ren91]	– Technical test
	– Ergonomics assessment:
	  – User trial
	  – Expert appraisal
	  – Performance test

Test environment	– Field study – Laboratory conditions
Measuring tool	– Subjective (questionnaire, observer, …) – Objective (EMG, interaction forces, …)

2.2 Types of Test

Wearable aids can be evaluated in technical tests or by ergonomics assessments. Technical tests examine the characteristics of a product and can verify the functionality of the aid. Technical tests can therefore assess some fundamental aspects of user acceptance, like efficiency and reliability. Nevertheless, technical tests are not able to evaluate all aspects of user acceptability.

Ergonomic tests ensure that the products are safe and convenient when in use. There are three types of tests which are the user trial, the expert appraisal and the performance test [Ren91].

In a user trial, potential or actual users are asked to use the product for a certain length of time. The sample of test persons should be drawn from the population most likely to use the product [Ren91]. For wearable aids, this is usually a group of patients with specific disabilities or people doing similar tasks at work. In [Ren91] the author claims that it is generally more effective to examine a few selected users in-depth instead of studying large samples in less detail. As social influences affect the user acceptability [Ven03], characteristics like age, sex and previous experience level should be considered. Since this chapter focuses on wearable aids with an intensive human-machine interface, anthropometric dimensions of the test persons have to be taken into account, too.

A test scenario independent of the user is the expert appraisal, in which experts evaluate the product using their experience and knowledge. The implementation of this test type is usually less complex than a user trial and therefore more feasible at less costs.

Performance tests simulate the use of the product without actually involving human subjects. This test type is appropriate when there may otherwise be a risk to the user. Within a performance test, usually physical measurements are conducted and connected to data on human abilities and limitations, e.g., using simulations. An example of such a test using a humanoid robot is shown in [Miu13, Ima14].

2.3 Test Environment

The evaluation can be either conducted in a natural environment (field study) or under laboratory conditions. While simple products or single features of a product can be tested under laboratory conditions within a short time, more complex products should be tested by the user over a longer period of time and under various situations, which is usually not feasible in a laboratory. Field trials are usually less controlled and difficult to observe but offer a realistic interaction [Ren91].

2.4 Measuring Tool

Depending on the measuring tool, a subjective or objective study is conducted. It is also possible to use several measuring tools within one study and therefore to combine subjective and objective methods.

Subjective tools usually rely on personal opinions and statements. The most common subjective measuring tool is a questionnaire [e.g., Bir17, Lob16]. The type of data that is required (norminal, ordinal, interval, ratio) influences the choice of scale for the questionnaire (e.g., dichotomous scale, rating scale, semantic differential scale). Furthermore, observers and video recording can be used to analyze the user's interaction with the wearable aid.

Objective studies are based on the observation of measurable facts. A common instrument to show that a wearable aid reduces the user's physical effort is the EMG measurement [e.g., Dij11, Saw08]. In order to prove that an exoskeleton reduces the wearer's metabolic cost, O_2 consumption and CO_2 production are measured using a gas analysis of the breathing air [e.g., Gal17, Seo16]. To assess the physical human-machine-interface the pressure distribution [e.g., Ros11, Bel08] or interaction forces [e.g., Rat16, Sch08] can be measured.

3 Case Study: Evaluation of a Passive Force Assisting Suit for Care Givers

Within this section, an exemplary evaluation of wearable aids is presented. This study aims to compare two force assisting suits and evaluate their user acceptance at a nursing home (user trial). It should reveal some requirements for practical use and show important aspects for developers of wearable aids for nurses. We assessed the subjective perception of the nurses using questionnaires.

3.1   Motivation

Nurses show an increased risk for musculoskeletal complaints [Eng96]. Especially in geriatric nursing low back neuromuscular fatigue after work was shown and nurses judged patient lifting, transfer, and turning as most physically demanding tasks [Hui01]. Tools like patient lifts, transfer boards, sliding aids and belts for support can reduce the physical work load. Their rare application is often justified by additional expenditure of time. Wearable aids offer an alternative, without the need of being carried to the operation site.

This study therefore evaluates the user acceptance of two Japanese passive force assisting suits in a German nursing home. Both suits are based on the same operation principle: Elastic belts over torso and buttocks are stretched while bending the upper body forward. Thus, the necessary force for performing actions in this body posture and for raising the torso should be reduced.

3.2   Method

Within this study, suit A and B have been worn during usual care work for one shift by 8 and 6 geriatric nurses, respectively. Therefore, this evaluation was a field study, as a natural setting was chosen. A quasi-experimental design was realized, because a natural group (caregivers at one nursing home) was investigated and no randomization in regard to the test persons was possible. The composition of participants is shown in Table 2. Because 84% of German nurses in patient care are female [Des15], this group allows conclusions for the relevant occupational group. Although this evaluation is based on a small sample, this first systematic survey allows an estimation of the user acceptance and reveals through the application in the usual work environment some practical requirements for such force assisting suits.

Table 2 Composition of participants with sex and age

		Suit A	Suit B
Sex	Male	0	0
	Female	8	6
Age	25–29	1	1
	30–39	3	2
	40–49	2	1
	50–59	2	1
	60–65	0	1

3.2.1 Test Objects

For one of the suits (suit A, a and b in Fig. 1) several studies conducted by the developers themselves concerning the effectiveness of the aid are published. They include EMG measurements [Ima11a], assessment of the postural stabilization [Ima13], field studies with nurses and questionnaires [Ima11a, Ima11b], and the evaluation with a humanoid robot [Miu13, Ima14]. About the second suit (suit B, c and d in Fig. 1) only some basic manufacturer specifications were known before the evaluation.

3.2.2 Measurement Instrument Questionnaire

Using a questionnaire, the subjective perception in quantitative parameters concerning the *support effect* and *comfort* as significant impacts on the user acceptance were assessed. In order to specify those two aspects, the participants could give further statements. Regarding the support effect, participants were invited to state if their muscle fatigue was reduced and whether an upright posture was assisted. In order to rate the comfort, they could state if pressure marks exist and if the suit obtains the full moving space. Finally, the participants were asked if they would *use the suit again*.

All test persons filled in the questionnaire prompt after wearing the suit. The questions were answered with visual analog scales from 0 to 100%, whereby 0% corresponded to a limitation and 100% corresponded to a benefit. The scales were read in 5%-intervals.

(a)   **(b)**   **(c)**   **(d)**

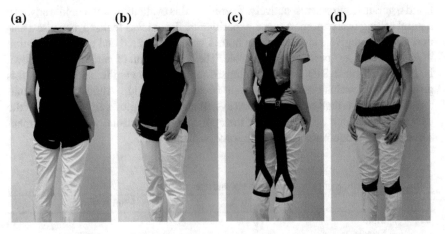

Fig. 1 Suit A (**a** and **b**): crossed belts beneath the vest are stretched when bending the torso forward. Suit B (**c** and **d**): crossed belts on the back and belts behind the thighs are stretched when bending the torso and knees

3.3 Analysis and Results

The assessment by nurses in the form of 8 questionnaires of suit A and of 6 questionnaires of suit B were analyzed. Since there was no normal distribution and only small sample sizes, we evaluated the data by means of characteristic values like median, median absolute deviation (MAD) and range. In order to visualize this data, boxplots were used (see Fig. 2).

We also showed a correlation between the subjectively felt comfort and the willingness to use the suit again. A scatter diagram (see Fig. 3) indicates a correlation between the two factors. We confirmed this relation by a hypothesis test using the Spearman's rank correlation coefficient.

4 Summary and Conclusion

In order to ensure the application of wearable aids their user acceptance has to be on a high level. For the evaluation of this key factor a classification regarding the target, type, environment and measurement tool is presented which aims to help developers and researchers as well as potential users to design assessments and rank existing studies. Within a case study the evaluation of two different force assisting passive suits are presented with the aim to show exemplary important steps of an assessment.

It was shown that the extent of an evaluation regarding wearable aids can vary considerably and has to be chosen carefully considering the aim of the study. The results of an evaluation help developers and users to guide their future work on the product by the identification of limitations and strengths. Moreover, a systematic evaluation supports users to identify suitable products meeting the needs of their application.

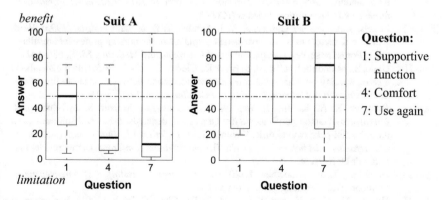

Fig. 2 Boxplots of answers for key questions for Suit A and B: the visualization shows more favorable valuation of Suit B

Fig. 3 The scatter plot of
the felt comfort and the
willingness to use the suit
again indicates a positive
correlation

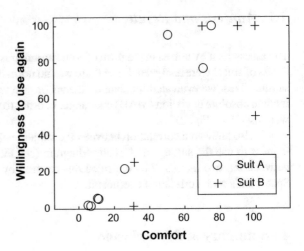

References

[Bel08]   Belda-Lois, J. M., Poveda, R., & Vivas, M. J. (2008). Case study: Analysis of pressure
           distribution and tolerance areas for wearable robots. In Pons, J. L. (Ed.), *Wearable
           robots: Biomechatronic Exoskeletons*. Wiley: Chichester.
[Bir17]   Birch, N., Graham, J., Priestley, T., Priestley, T., Heywood, C., Sakel, M., et al. (2017).
           Results of the first interim analysis of the RAPPER II trial in patients with spinal cord
           injury: Ambulation and functional exercise programs in the REX powered walking aid.
           Journal of Neuroengineering and Rehabilitation, 14(1), 1–10.
[Dav89]   Davis, F. D. (1989). Perceived usefulness, perceived ease of use, and user acceptance of
           information technology. *MIS Quarterly, 13*(3), 319–340.
[Des15]   Destatis. (2015). Pflegestatistik 2013: Pflege im Rahmen der Pflegeversicherung -
           Deutschlandergebnisse. Statistisches Bundesamt, Wiesbaden.
[Dij11]   van Dijk, W., van der Kooij, H., & Hekman, E. (2011). A passive exoskeleton with arti-
           ficial tendons: Design and experimental evaluation. In *2011 IEEE International Con-
           ference on Rehabilitation Robotics (ICORR)*.
[Eng96]   Engels, J. A., van der Gulden, J. W. J., Senden, T. F., & van't Hof, B. (1996). Work
           related risk factors for musculoskeletal complaints in the nursing profession: results of
           a questionnaire survey. *Occupational and Environmental Medicine, 53*(9), 636–641.
[Gal17]   Galle, S., Malcolm, P., Collins, S. H., & De Clercq, D. (2017). Reducing the metabolic
           cost of walking with an ankle exoskeleton: Interaction between actuation timing and
           power. *Journal of Neuroengineering and Rehabilitation, 14*(1), 1–16.
[Goe16]   Goehlich, R. A., Krohne, I., Weidner, R., Gimenez, C., Mehler, S., & Isenberg, R.
           (2016). Exoskeleton portfolio matrix: Organizing demands, needs and solutions from
           an industrial perspective. In R. Weidner (Ed.), *Technische Unterstützungssysteme, die
           die Menschen wirklich wollen: Zweite Transdisziplinäre Konferenz*. Hamburg: Helmut-
           Schmidt-Universität.
[Ham13]   Hambling, B., & van Goethem, P. (2013). *User acceptance testing: A step-by-step guide*.
           Swindon: BCS Learning & Development.
[Hui01]   Hui, L., Ng, G. Y. F., Yeung, S. S. M., & Hui-Chan, C. W. Y. (2001). Evaluation of
           physiological work demands and low back neuromuscular fatigue on nurses working in
           geriatric wards. *Applied Ergonomics, 32*(5), 479–483.

[Ima11a] Imamura, Y., Tanaka, T., Suzuki, Y., Takizawa, K., & Yamanaka, M. (2011). Motion-based-design of elastic material for passive assistive device using musculoskeletal model. *Journal of Robotics and Mechatronics, 23*(6), 978–990.

[Ima11b] Imamura, Y., Tanaka, T., Suzuki, Y., Takizawa, K., & Yamanakam, M. (2011). Motion-based design of elastic belts for passive assistive device using musculoskeletal model. In *2011 IEEE International Conference on Robotics and Biomimetics (ROBIO)*.

[Ima13]   Imamura, Y., Tanaka, T., Nara, H., Suzuki, Y., Takizawa, K., & Yamanaka, M. (2013). Postural stabilization by trunk tightening force generated by passive power-assist device. In *2013 35th Annual International Conference of the IEEE Engineering in Medicine and Biology Society (EMBC)*.

[Ima14]   Imamura, Y., Tanaka, T., Ayusawa, K., et al. (2014). Verification of assistive effect generated by passive power-assist device using humanoid robot. In *2014 IEEE/SICE International Symposium on System Integration (SII)*.

[Lob16]   Lobo-Prat, J., Kooren, P. N., Janssen, M. M. H. P., Keemink, A. Q. L., Veltink, P. H., Stienen, A. H. A., et al. (2016). Implementation of EMG- and force-based control interfaces in active elbow supports for men with duchenne muscular dystrophy: A feasibility study. *IEEE Transactions on Neural Systems and Rehabilitation Engineering, 24*(11), 1179–1190.

[McC05]   McCreadie, C., & Tinker, A. (2005). The acceptability of assistive technology to older people. *Ageing & Society, 25*(01), 91–110.

[Miu13]   Miura, K., Yoshida, E., Kobayashi, Y., et al. (2013). Humanoid robot as an evaluator of assistive devices. In *2013 IEEE International Conference on Robotics and Automation (ICRA)*.

[Rat16]   Rathore, A., Wilcox, M., Ramirez, D. Z. M., Loureiro, R., & Carlson, T. (2016). Quantifying the human-robot interaction forces between a lower limb exoskeleton and healthy users. In *2016 38th Annual International Conference of the IEEE Engineering in Medicine and Biology Society (EMBC)*.

[Ren91]   Rennie, A. M. (1991). The application of ergonomics to consumer product evaluation. *Applied Ergonomics, 12*(3), 163–168.

[Ros11]   de Rossi, S. M. M., Vitiello, N., Lenzi, T., Ronsse, R., Koopman, B., Persichetti, A., et al. (2011). Sensing pressure distribution on a lower-limb exoskeleton physical human-machine interface. *Sensors, 11*(1), 207–227.

[Saw08]   Sawicki, G. S., & Ferris, D. P. (2008). Mechanics and energetics of level walking with powered ankle exoskeletons. *Journal of Experimental Biology, 211*, 1402–1413.

[Sch08]   Schiele, A. (2008). Case study: Quantification of constraint displacements and inter-action forces in nonergonomic pHR interfaces. In Pons, J. L (Ed.), *Wearable robots: Biomechatronic exoskeletons* (pp. 149–154). Wiley: Chichester.

[Seo16]   Seo, K., Lee, J., Lee, Y., Ha, T., & Shim, Y. (2016). Fully autonomous hip exoskeleton saves metabolic cost of walking. In *2016 IEEE International Conference on Robotics and Automation (ICRA)*. IEEE.

[Ven91]   Venkatesh, V., Morris, M. G., Davis, G. B., & Davis, F. D. (1991.) User acceptance of information technology: Toward a unified view. *MIS Quarterly, 27*(3), 425–287.

[Ven03]   Venkatesh, V., Morris, M. G., Davis, G. B., & Davis, F. D. (2003). User acceptance of information technology: Toward a unified view. *MIS Quarterly, 27*(3), 425–287.

Extended Model for Ethical Evaluation

Karsten Weber

Abstract Failure to pay attention to ethical, legal, and social aspects or impacts (ELSA/ELSI) of technology can have considerable negative repercussions like lack of acceptance of a new technology, product, or service among prospective users and thus economic failure. Furthermore, funding institutions expect researchers and developers taking ELSA into account; in the EU, this is called Responsible Research & Innovation. In order to implement this concept in R&D, tools are needed; in what follows MEESTAR and its successor MEESTAR^2 shall be presented. Initially developed for the ethical evaluation of ambient assisted living systems, MEESTAR^2 can also be used to evaluate other technologies.

1 Introduction

For some years now and particularly in the context of publicly funded basic research as well as in funded R&D it has increasingly been emphasized that the development of new technology, products, and services must not only focus on technology itself, but that its impact on individuals, social groups, societies, or the environment must also be investigated in advance in order to minimize or even prevent negative repercussions and strengthen positive effects. This requirement originates from the field of technology assessment; today it is referred to as considering ethical, legal, and social aspects or impacts of technology—in short ELSA or ELSI. The European Commission speaks about "Responsible Research & Innovation" (RRI) [EUC12] and calls for this concept to be taken into account in all EU-funded projects regardless of the thematic scope.

There are different views on the meaning of RRI [Bur17] and there are just as many tools to integrate ethical considerations into development of technology [Rei17]. The following is a tool that was originally developed on behalf of the German Federal

K. Weber (✉)
Institute for Social Research and Technology Assessment (IST), Ostbayerische Technische Hochschule (OTH), Regensburg, Germany
e-mail: Karsten.Weber@oth-regensburg.de

© Springer Nature Switzerland AG 2018   257
A. Karafillidis and R. Weidner (eds.), *Developing Support Technologies*, Biosystems & Biorobotics 23, https://doi.org/10.1007/978-3-030-01836-8_25

Ministry of Education and Research (BMBF) for the ethical evaluation of ambient assisted living (AAL) systems. However, MEESTAR can also be used in other contexts if appropriate adjustments are made.

In what follows, the history of MEESTAR's development is briefly described, then the link to other ethical evaluation tools is shown and finally, it is illustrated how MEESTAR[2] as a follow-up can be used with regard to technology other than AAL.

2 Demographic Change, Care, and Technology

In response to the demographic change that takes place not only in Germany but in many other industrialized countries, it has been discussed for some time to massively employ technology in healthcare to meet the societal challenges caused by this change and to contribute at least to the following aims: (a) Technology shall contribute to reduce costs in healthcare since demographic change could lead to considerable funding gaps in social security systems and also raise questions of intergenerational justice. (b) Since already today healthcare providers have considerable difficulties in meeting their labor requirements, technology also shall help preventing labor shortages. (c) Additionally, it is intended to help nurses as well as informal caregivers in carrying out stressful activities to prevent serious health problems on their side. (d) Technology shall help to provide healthcare in rural and remote areas with inadequate workforce. (e) Last but not least, age-appropriate assistive systems are supposed to help open up new markets and thus promote economic prosperity.

Age-appropriate assistive or AAL systems cover, for example, computer games designed to help maintain mental fitness and performance, especially for the elderly, tele-monitoring and telecare systems to provide healthcare particularly in rural and remote areas, social robots such as Paro, and highly networked systems as well as service and household robots that help people with physical and mental handicaps to live (more) independently (cf. [Ras13]). Although this list is certainly not exhaustive, it already shows the diversity of AAL systems; it makes also clear that all these systems focus on human interaction. This raises questions that are also relevant in relation to other technologies and already suggests that a tool for the ethical evaluation of AAL systems probably could also be used in other contexts: For instance, with regard to cyber-physical systems it is very likely that the deployment of massively interconnected systems will increase productivity, but at the same time exert increased performance pressure on employees through the collection of behavioral data; the use of collaborative robots for human-technology interaction in production processes could lead to competition between safety and usability; the benefits of new digital production methods could be unfairly distributed among companies and their employees—this rather short list of examples already shows that normative conflicts can surely be found in industrial environments. Therefore, it makes sense to look for ways of minimizing or completely preventing such conflicts in advance. Although initially developed to only be used in the context of AAL systems MEESTAR is a tool that can help with identifying normative conflicts and with the development of possible solutions.

3 MEESTAR

No matter how promising AAL systems seem to be, the use of technology aimed at supporting elderly persons raises far-reaching social and normative questions. This makes it necessary to evaluate possible effects of technology on stakeholders as well as on moral norms and values not only before (or even after) its deployment, but also before and during its development. Among other things, this includes a critical look at leitmotifs of AAL systems: Information presented at trade fairs, conferences, or on the Internet most often show seniors who are seemingly in a dazzling physical and psychological state. But even nowadays age is not only defined by grey haired but still attractive women and men, but rather often means illness, infirmity, suffering, and complete dependence on the help of others—elderly persons belong to a group of people who are vulnerable in many ways. A central task of applied ethics is to challenge biased images and role models of age, to hint at the diversity of existing and possible ways of life and to communicate both positive and negative aspects of old age.

The discrepancy between idealized and realistic images of age and ageing may have been one reason for the BMBF to recognize the need for research concerning normative questions raised by AAL systems and its commissioning of a project for the ethical evaluation of this technology in 2012 [Man15]. Main objectives of that project were to identify relevant moral values (called 'dimensions') of AAL systems, the development of an ethical evaluation tool, and the formulation of (fifteen) guidelines aimed at all relevant stakeholders in the context of AAL systems like system developers, patients and users, nurses, healthcare providers, and relatives. The tool being developed in this project is called MEESTAR for 'Model for the ethical evaluation of socio-technical arrangements' which primary objective is to generate awareness of moral problems on the part of as many stakeholders as possible. However, it has often been shown in past applications that MEESTAR can also contribute to look at non-ethical aspects of AAL systems from rather unexpected perspectives, as stakeholders who are otherwise often not consulted during the design and development process get a voice.

It must be emphasized that the use of MEESTAR cannot and must not be aimed at unconditional acceptance or fundamental rejection of technology, but must always intended to answer situation-dependent considerations like: Are we confronted with a situation in which technology is really the solution? Would a high-tech or low-tech version make more sense? Should permanent or temporary solutions be sought? Do we really need all the data the system shall gather or can we reduce the amount of data? MEESTAR shall make trade-offs of each considered alternative explicit and visible to all stakeholders in order to enable them to assess benefits and burdens, opportunities and risks, advantages and disadvantages, profits and costs of technology on the basis of reliable information. For responsible decisions can only be made by stakeholders on a well-informed basis. The aim should be to ensure that an ethical evaluation, contrary to the classic waterfall model of technology development, is carried out not only once and then most likely at the end of the development process, or even as late as when companies push their products into the market, but already during the design and development phases of these products and in the best case iteratively, in order to receive feedback time and again on how the product fits the expectations of stakeholders.

4   Extended MEESTAR: MEESTAR^2

One important methodical challenge for MEESTAR that must not be neglected is that it might influence the involved stakeholders in their judgments; in this regard this tool is no different from other (social science) survey methods. Proposing certain moral values represented by the so-called 'dimensions' harbors considerable potential for influencing the evaluation process since, among other aspects, participants could not take other moral values into account and are confronted with preconceptions concerning the meaning of those moral values. Even though this can only be regarded as an anecdotal evidence so far, it should be noted that the use of MEESTAR in educational research projects as well as in the evaluation of R&D projects has repeatedly shown that, from the point of view of the stakeholders involved, the selection of the normative dimensions does not prove to be self-evident, appears to be worthy of discussion and requires a more systematic justification than has been available so far. All of these facts suggest that MEESTAR should be further developed, e.g. in order to better justify the normative dimensions or revise their selection.

An extended version of MEESTAR could be based on a number of established approaches and methods in the field of ethics of technology, Health Technology Assessment, and participatory technology design. In this regard, particularly the approach of Value Sensitive Design (VSD) has to be mentioned which was propagated by Batya Friedman [Fri96] and explicitly aims at recognizing those moral values (what in MEESTAR is referred to as 'dimensions') that could be affected by a particular design of technology. VSD's methodology can thus be used to systematically identify moral dimensions for MEESTAR, which can then be ethically evaluated. This entails additional efforts, which, in view of the generally rather scarce resources devoted to ethical, legal, and social aspects of technology, will often not be easy to achieve. One way of minimizing this problem could be to search the extensive literature on VSD in the context of various technologies to find out which relevant standards and values have already been identified there.

A certain similarity of MEESTAR can also be found in relation to the Ethical Matrix [Mep00]. This method is intended to use the four basic principles of Beauchamp and Childress [Bea06] (autonomy, beneficence, non-maleficence, and justice) to evaluate the consequences of the use of technology with regard to the promotion or violation of these principles. It is obvious that these can be translated almost entirely directly into MEESTAR dimensions, such as in the case of self-determination and privacy, which fall under autonomy, care and participation under beneficence, security under non-maleficence, and justice can be translated directly; only the dimension of self-perception cannot be readily translated into one of the four principles. Indeed, self-perception is the one dimension that has provoked the most debates in MEESTAR workshops held so far, be it in student or R&D projects. Actually, this dimension is intended to stimulate the MEESTAR participants to reflect, for instance, on possible changes in their own professions through technology and to evaluate these changes from a moral point of view.

Although the above-mentioned principles were initially developed for the medical field, they have become the basis for a variety of ethical evaluation methods [Rei17]: Security has high priority especially in industrial production processes, where data protection and privacy are also playing an increasingly important role. However, undoubtedly not for all MEESTAR dimensions an equivalent can be found in industrial contexts—a basic idea of MEESTAR^2 is that the relevant normative dimensions are also negotiated among stakeholders and only then does the actual evaluation begin. For instance, it is very likely that usability will be an important dimension. Actually, the dimensions must not be 'normative' in a narrow sense; they must only meet the condition to be of value for the stakeholders. If the evaluation then reveals that the technology under investigation does not adequately support this dimension, the demands of the stakeholders cannot be implemented appropriately; a possible consequence of this is a lack of acceptance (Fig. 1).

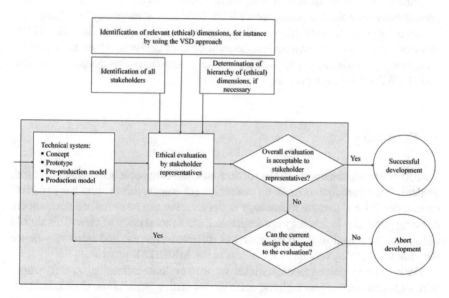

Fig. 1 Flow chart of MEESTAR^2

5   Necessary Further Steps

There is indeed a close connection between normative technology assessment on the one hand and participatory technology design and acceptance of technology on the other. For technology acceptance research the inclusion of prospective users is methodologically indispensable in order to establish acceptance or rejection. It is irrelevant whether a simple or a more sophisticated technology acceptance model is used—in all models user preferences have to be determined. If the results of these analyses are then taken into account in the design of technology, one has come very close to participatory technology design. In fact, MEESTAR^2 can be seen as a part of a requirements engineering including moral requirements. With regard to engineers, developers, designers and the like it seems to be a good idea to "sell" MEESTAR^2 primarily not as an ethical evaluation tool but as a method to evaluate user or, better to say, stakeholder requirements.

There is at least one more important desideratum: From the point of view of those who are to design and develop technology, it would probably be reasonable if (extended) MEESTAR would prioritize moral dimensions. Up to now, these have been on an equal footing; the original idea behind it was that the employment of MEESTAR is primarily aimed at the development of problem-solving proposals to mitigate or remedy moral concerns with regard to technology. However, it is quite conceivable and even very likely that moral values are in competition with each other. In such a situation, prioritization, which also might be necessary due to legal requirements like industrial safety or privacy legislation and/or other laws, would certainly be helpful concerning the design of a product and/or service. If prioritization is also carried out within the framework of a MEESTAR^2 run, if necessary by taking legal requirements into account, this would also have the advantage that the stakeholders involved would have to give mutual account of their decisions about their choices concerning prioritization. The justifications and arguments thus gained might also be helpful for the design process.

6   Summary

If MEESTAR is understood as a kind of discourse ethics approach, its benefit becomes evident: If the participants make an evaluation (of course, this process has to satisfy certain conditions) that the technology under scrutiny can be described normatively as being good, this technology is acceptable from a moral point of view. This should always be done against the background of different expertise and interests—hence the aim of involving various stakeholders in the MEESTAR process.

For economic reasons alone, it makes sense to evaluate technology at early stages of design and development with respect to the normative expectations of stakeholders. Thus, it belongs to the context of participatory or constructive technology assessment. This not only represents an ethical demand in the sense of empowerment, but also a method to ensure acceptability and usability. It also promotes value-based design of technology and thus follows the EU's approach of Responsible Research and Innovation.

References

[Bea06]   Beauchamp, T.L., & Childress, J. F. (2006). *Principles of biomedical ethics* (6th ed.). Oxford University Press.

[Bur17]   Burget, M., Bardone, E., & Pedaste, M. (2017). Definitions and Conceptual dimensions of responsible research and innovation: A literature review. *Science and Engineering Ethics, 23,* 1–19.

[EUC12] EU Commission: Responsible Research and Innovation. Europe's ability to respond to societal challenges, 2012. Available at https://ec.europa.eu/research/swafs/pdf/pub_rri/ KI0214595ENC.pdf, last visited January 10, 2017.

[Fri96]   Friedman, B. (1996). Value-sensitive design. *Interactions, 3,* 16–23.

[Man15] Manzeschke, A., Weber, K., Rother, E., & Fangerau, H. (2015). Ethical questions in the area of age appropriate assisting systems. VDI/VDE.

[Mep00] Mepham, B. (2000). A framework for the ethical analysis of novel foods: The ethical matrix. *Journal of Agricultural and Environmental Ethics, 12,* 165–176.

[Ras13]   Rashidi, P., & Mihailidis, A. (2013). A survey on ambient-assisted living tools for older adults. *IEEE Journal of Biomedical and Health Informatics, 17,* 579–590.

[Rei17]   Reijers, W., Wright, D., Brey, P., Weber, K., Rodrigues, R., O'Sullivan, D., & Gordijn, B. (2017). Methods for practising ethics in research and innovation: A literature review, critical analysis and recommendations. *Science and Engineering Ethics* (online) December 9, 2017.

Legal Responsibility in the Case of Robotics

Susanne Beck

Abstract The development of robotics poses problems in terms of ascribing responsibility to specific individuals. This could leave the person whose rights are violated by a robot without possibility to receive damages or, in general, making someone legally responsible for the damage done. Several possible solutions to these problems posed by the technological developments are discussed, such as the introduction of the so-called "electronic person". But these solutions will have repercussions on social concepts such as personhood, dignity, or responsibility. The article will analyze some of the legal problems posed by robotics, will show and discuss some of the discussed solutions and their possibleconsequences.

1 Introduction

Robotics is one of the most important technologies of our time and will become more relevant in many areas of life in the years to come. At least in some contexts, machines will become more and more "autonomous"[1] by approximating human thought patterns [Wen09]. These machines will facilitate everyday life, extend the

---

[1] In the following, "autonomous" is used in a broad sense, meaning nothing more than a certain space for decision-making for the machine. For a project working on different understandings of autonomy and their changes because of new human-technology interactions see http://www.isi.fraunhofer.de/isi-de/v/projekte/WAK-MTI.php. Of course, notions such as "decision" or "learning", which I will have to use at some points in this paper, are not meant to imply that these processes are similar to human processes of deciding or learning—instead, these are always insufficient analogies (see to the criticism about the similarity argument below). Still, until now, there are no more adequate notions to describe the processes, and the analogies do represent the minimum resemblance between humans and machines which is, inter alia, problematic.

This article is a short version of the former publication "The problem of ascribing legal responsibility in the case of robotics", AI & SOCIETY 2015.

S. Beck (✉)
Faculty of Law, Leibniz University Hanover, Hanover, Germany
e-mail: susanne.beck@jura.uni-hannover.de

A. Karafillidis and R. Weidner (eds.), *Developing Support Technologies*, Biosystems & Biorobotics 23, https://doi.org/10.1007/978-3-030-01836-8_26

time, the elderly can live at home, and might, to a certain extent, even replace social (human) contacts and fulfil emotional needs. They will relieve humans of specific tasks and decisions, and it is possible that for some tasks the decision of a machine might even be quicker, more rational, and more informed than a human decision.[2] Like any other technological development, the progress of robotics will pose new challenges for the normative systems of society, morality, and the law, e.g., ascribing responsibility will become more difficult because of the autonomy of the machines. Regulations for human behaviour—including the programming and use of these machines—will have to be developed and existing norms will have to be adapted. In the following, I will give an overview over the problems with ascribing legal responsibility that occurred due to the recent developments in robotics and discuss some possible solutions. The focus will lie on tort law and product liability as one of the main actual challenges for robotics and as striking example for the problems and possible solutions of legal responsibility ascription.

2   Responsibility Ascription for Damages

The recent developments in robotics pose enormous difficulties for the legal system. While—as we will see in the following—robotics is a collaborative project where some responsibility is even transferred onto a machine, the legal system is, traditionally, oriented towards individual responsibility and therefore has difficulties to cope with this technology.

Some of the responsibility ascription problems posed by robotics can be shown by trying to answer the—in this context often discussed—question about the legal consequences of an autonomous robot damaging a third party.[3]

2.1   The Robot Causing Damage Because of Manufacturing Defects

One possible scenario is that the damage[4] has been—provably—caused by manufacturing defects.[5] In this case, the responsibility ascription does not pose any additional problems in the case of robotics, as can, e.g., be shown on an EU-level: The Product

---

[2] "Fortunately, these potential failings of man [*passion for inflicting harm, cruel thirst for vengeance, unpacific and relentless spirit, fever of revolt, lust of power, etc.*] need not be replicated in autonomous battlefiled robots" [Ark08, p. 2].

[3] The following aspects are discussed in more detail in [EuR13].

[4] For the purposes of the Product Liability Directive 85/374, "damage" means (Article 9) damage caused by death or by personal injuries; damage to an item of property intended for private use or consumption other than the defective product itself.

[5] Article 6 product liability Directive 85/374 states that a product is defective when "*it does not provide the safety which a person is entitled to expect, taking all circumstances into account,*

Liability Directive 85/374 imposes the principle of strict liability (liability without culpability) on the producer if the damage was caused by a defective product. If more than one person (manufacturer, supplier, or importer) is liable for the same damage, the consequence is joint liability.

The main challenge of this scenario is that the burden of proof for the actual damage, the defect in the product, and the causal link between damage and defect lies with the wronged party—no proof is necessary for the negligent or faulty behavior on the part of the producer or importer, though, because of their strict liability. Strict liability is based on a decision by the lawmaker that one party is responsible for a specific damage even if the social inadequacy of the action cannot be proven. The reason for this decision is, usually, that this party is the one profiting the most, controlling the process, and in most cases actually causing the damage by some mistake. But even the proof of the defect itself could be difficult in the case of robotics, especially if the computer program is complex and not readable ex post.

Under specific circumstances, the producer can excuse himself from strict liability,[6] for example, if he can prove (Art. 7): *"(e) that the state of scientific and technical knowledge at the time when he put the product into circulation was not such as to enable the existence of the defect to be discovered; or (f) in the case of a manufacturer of a component, that the defect is attributable to the design of the product in which the component has been fitted or to the instructions given by the manufacturer of the product"*.

Both cases are likely to be relevant for robots. Because of the strong research activity, it is probable that new findings show deficiencies of former versions of AI-programs—thus also showing that the producer could not have acted differently. And because in most cases different manufacturers (and programmers) contribute to a robot, it also is quite possible that one of the manufacturers of a component can show that the design or instructions lead to defects thus freeing himself from liability and transferring it to the designer or instruction giver. It is difficult to draw an exact line for the possibility of manufacturers to refer to the state-of-the-art defence just yet—for this one we will have to await jurisprudence on the topic of autonomous machines. But one has to keep in mind that the people researching, producing, and using robots know about the uncertainties of these machines.

---

including:(a) the presentation of the product;(b) the use to which it could reasonably be expected that the product would be put;(c) the time when the product was put into circulation".

[6]The producer is freed from all liability if he proves (Article 7): *"(a) that he did not put the product into circulation; or (b) that, having regard to the circumstances, it is probable that the defect which caused the damage did not exist at the time when the product was put into circulation by him or that this defect came into being afterwards; or (c) that the product was neither manufactured by him for sale or any form of distribution for economic purpose nor manufactured or distributed by him in the course of his business; or (d) that the defect is due to compliance of the product with mandatory regulations issued by the public authorities; or (e) that the state of scientific and technical knowledge at the time when he put the product into circulation was not such as to enable the existence of the defect to be discovered; or (f) in the case of a manufacturer of a component, that the defect is attributable to the design of the product in which the component has been fitted or to the instructions given by the manufacturer of the product"*.

Although these aspects challenge the attribution of existing law in a new area, still, a conscious interpretation will probably be able to solve these problems and thus adapt the laws to the characteristics of robots as products. One has to be aware of the difficulties to prove the mistake in a program as complex as used for "autonomous" robots, though. Thus, it can be stated that the more complex the robot is, the more autonomous the machine acts, and the more difficult it becomes for the wronged party to prove the link between a specific mistake of one of the parties involved and the damage.

2.2   Robots as Products? Mistakes as Defects?

When adapting the existing laws, one could already question if the premise to treat robots as mere products is convincing. This premise is challengeable by the fact that soon robots could be equipped with an adaptive and learning ability. These features necessarily involve a certain degree of unpredictability and uncontrollability in the robots' behavior: Because of the increase in experience gained by the robot on its own, the robot's conduct cannot be planned in its entirety anymore.

This does not only lead to doubts about the product-quality of robots, but also to the question whether every "mistake" by the robot is necessarily caused by a defect in the legal sense [Bos11]. If robots with adaptive and learning capabilities[7] are left free to interact with humans in a non-supervised environment, they could react to new inputs they received in an unpredictable way. If a robot then causes damage to a third party because of these reactions it is hardly plausible that the robot was defect; it did what it was supposed to do: It reacted to new inputs and adapted its behavior—thus the machine is not defective as such.

Currently, there are no tort regulations for such cases. The creation of an analogy to the regulations for owners of animals is sometimes discussed; an argument for such an analogy could be that the underlying conflict is adequately similar—in both cases, the actions of animal/machine are partly unpredictable but can be influenced by training and control. Also, the limits between objects and animals—mainly the ability to move freely in the surrounding space—have been overcome by robots. One could even think about assuming the parental model, comparing cognitive robots to children who learn during their own path of growth. Minors act due to their upbringing and should be guided by their parents, robots act due to the behaviour they were taught and must be instructed by the user. More abstractly, one can surely set certain guidelines for dealing with these kinds of machines: The higher the robot's capability to learn, the lower the producer's responsibility; the longer the duration of a robot's instruction, the greater the responsibility of its owner. This only accounts

---

[7]First of all, again, the notions of adaptation and learning are meant only as analogies to human processes. Second, it is important to note that, although in many ways the problem of "many hands" [Jon82] is relevant here, the possibility of learning and adaption is, in fact, the crucial difference. Because this development is inbuilt into the machine, one does not only deal with side-effects of cooperation but with intended effects of uncontrolled development of machines.

for acquired abilities (resulting from instruction) which therefore should be set apart strictly from programmed characteristics not changeable by learning or education.

Besides these guidelines one should think about possible legal concepts to solve the underlying conflict in a more abstract way (behind the idea of transferring the mentioned legal concepts onto the situation here one obviously also finds abstract principles).

The most relevant conceptual solutions to the problem that robots—in opposition to normal products—might be trained by the user, might act unpredictable, and might cause damage in specific situations without there being a defect or necessarily a misuse, are:

(1) One of the human parties is regarded as generally liable, e.g., the user. This is how, for example, the law handles parking distance control systems at the moment.[8] The solutions to apply laws concerning pet owners or parents are based on this idea as well. The reasoning behind this solution is that the user is the one training the robot, deciding to use it in a specific situation and relying on it (e.g., not controlling it sufficiently). On the other hand, this is not just problematic in situations in which the damage really was a defect—often it also contradicts the idea of using a machine to relieve the user if he has to control it.

(2) Only the human party is liable who, provably, made a mistake. This legal concept is based on the idea that one, generally, is only liable for the damage one provably has caused intentionally or negligently. The legal exceptions to this principle are, usually, made explicit and concerned with situations in which disproportionate social power has to be balanced. This solution would put the burden of proof onto the third (wronged) party, posing a strong hindrance to their likelihood of receiving damages. Therefore, this solution is probably the most questionable because in these cases the control over the machine, as well as the advantages of its usage, actually lie at least as much with all other parties involved (programmer, producer, user) who profit from the purchase as with the third party.

(3) The consequence of the latter aspect and the insufficiencies of traditional solutions could be that all human parties "behind" the robot (researcher, programmer, producer, user) are transformed into a new legal entity ("electronic person"—more on this concept later). This does not, per se, solve all problems and does not necessarily exclude other solutions but gives the third party a kind of addressee, at least for its financial claims. Thus, the concept is meant to disburden the damaged party of having to prove who of all the parties is actually responsible for the damage in question. Thereby financial recovery is ensured: It is, e.g., thinkable that the user of the robot is, compared to the producer, not able to recover all damages. In such cases, one of the other parties can be obliged to pay for the damages as well. One has to be aware, though, that this concept can serve as hide-out for the responsible individuals (as can, e.g., corporate liability). Thus, one has to ensure that this solution does not hinder choosing other legal ways, either for the third party or for the other participants,

---

[8]See Amtsgericht München, Urteil vom 19.7.2007 – Az.: 275 C 15658/07, NJW RR 2008, 40.

to adjudge the individual responsible if someone actually has made a mistake. It also has to be legally communicated that this possibility and financial deposit cannot justify the building of robots which endanger others. One way to communicate this legally could be to strengthen criminal sanctions in this area of life for (grossly) negligent behaviour and ensure the prosecutions in practice.

(4) One could even, e.g., for socially useful robots (e.g., garbage collecting robots), think about transferring part of the damages onto the society itself, meaning the tax payer would have to help out in the case in which the origin of the mistake cannot be proven. This is based on the argument that whoever profits from the usage of the machine should also be—at least partly—liable for any financial disadvantages of third parties. Again, as in the case of the electronic person, one should not only focus on this construct but still encourage those possibilities where individuals are made responsible for their actions and decisions if inadequately dangerous.

These are just some legal suggestions (mainly focusing on private law) which are not all equally persuasive but some of them could be combined in various ways. They all stem from the specific balancing of certain aspects relevant to the decision about (financial) responsibility for damage: causation, responsibility-sphere (control of production process; decision about usage), risk-predictability and social adequacy of risk, advantages of usage, etc. These are only some aspects showing the complexity of the responsibility question—although the question here is only what happens after an "autonomous" robot caused damage.

3   Responsibility for Risks and Side Effects

Besides asking the question of responsibility for damages, one also has to discuss another responsibility ascription problem: Who is responsible (prospectively) for the decision about the areas of life in which one actually should use robots?[9] Obviously, as for any new product, this is the decision of the single researcher who is interested in further developing specific capacities of the machines, as well as one of the producer who brings the product to market and also of the customers who are either interested in buying it or not. But it also is for other institutions to contribute to the decision, especially for research funds (by financing only specific research), insurance companies (by ensuring under certain conditions), and political institutions (by giving out licenses). Also, the legislator can either facilitate or hinder the introduction of robots

---

[9]This is not a typical legal question, thus it is not discussed in depth here—the background aspects would also probably require a different paper. It is mainly for politics (and ethics, political and social sciences) to discuss this question in more detail. Still, I think it should be mentioned because until now, the main focus of the debate lies on the question of responsibility after damaging a third party. For this to happen, one has to decide beforehand if robots should actually be used in a certain area of life. This question is rarely discussed—in public as well as in the academic debate—although it should, in my eyes, be the first step before discussing responsibility for damages. For further inspiration see, inter alia, [Bro08].

to certain areas of life, for example by deciding in favour of one of the mentioned solutions for the responsibility for damages (the less reparations the programmer and producer have to fear to have to pay out, the more likely the introduction to market will be). Again, this is the same for every new technology. Still, these decisions about the development of the society in this field create responsibilities—not just because of the high costs of the development of robots, but also because in many ways, these developments will be connected to aspects of social justice (health, care, mobility for elderly, education, etc.).

Finally, the question in which areas one can accept robots will depend on their potential for damage. Because even if one could solve the problem of financial compensation, not all immaterial damages can be fully compensated for. Thus, it might be regarded as inadequate to actually expose third parties to a risk in certain areas, if the social advantage of the robot is not high enough. Surely, this is not only a problem encountered by robotics but by all technologies which expose third parties to risks. Still, one has to be aware that it has to be discussed in detail who should make these decisions and how it can be ensured that they mirror the interests of the society.

This is important because "risk" in this context does not only mean the damage to third parties: There are also the risks of certain unwanted side effects. Again, it can be said that most new technologies are accompanied by doubts about possible slippery slopes (be it the loss of numerous work places because of robots, decreasing human contact in society, the usage of Autonomous Weapon Systems by terrorists). Still, it is not surprising and has to be taken seriously that these concerns are raised in the case of robotics: robots nursing the elderly or baby-sitting, taking over human communication, giving advice or waging wars will surely change our perception of social interactions and challenge our understanding of "social" [Fit13].

4 Responsibility in spite of Responsibility-Transfer

One idea of developing more and more autonomous machines is to transfer responsibility onto machines: Overwhelmed by the sometimes called "tyranny of choice" in complex situations in everyday life entailing endless and unforeseeable risks of damaging third parties, we nowadays tend to react with technology. We are building machines not just to decide how to find the best way in traffic or to get our car into a parking lot, not just to remind us to take our medication or to buy food as soon as the fridge is empty—we are even building machines to decide about life and death of other human beings. This development poses, in my opinion, a responsibility ascription problem per se: When machines take over decisions, new questions will arise not just if something goes wrong, but for each and every decision made by machines: Who is the responding entity? Can the machine respond in a way that is necessary for the social and legal construct of responsibility?

The problem of this responsibility transfer can be seen by looking at the potential social reactions to decisions of machines [Bot14]. To a certain extent, social problems could arise because of the differing perception of humans and machines. It seems plausible that it is, for most humans, harder to forgive a machine if it makes a decision which affects a human life—not just if this decision was wrong, but also, if a correct decision has negative effects, e.g., if a robot would tell an employee that because of the calculation of a computer program he loses his job (or write an email of which one knows it is written by a machine) or—coming back to the drastic demonstrative example of the military—if the robot decides without involving a human that a human life has to end. Because of society's doubts whether a machine can actually be the one responding to make such decisions, it is debated if it were a violation of human dignity to let a machine make this life-or-death decision [Hey13]. The argument goes that human dignity includes the right that a human being has to decide on the end of human lives, that a human being has to live with and "respond" to the consequences of this decision.

But the military is not the only area in which the necessity of human response could exclude the transfer of decision making onto machines, even if factually possible and sometimes even if the machine might make less mistakes than human beings. This necessity is also likely in—just to name a few examples—the medical area, education, care for the elderly, and the law. Human response being a social condition for certain decisions at the moment does not mean that this condition will lose importance when social concepts are changing (more on that in a minute). It will have to be discussed in every situation if this condition has to be respected for now, if it has already changed, what human response means in a specific context, etc.

5 Conclusion and Repercussions

As shown, the development of robotics does pose responsibility ascription problems not just in the case of robots causing damage to third parties. It also means that one has to remember one's responsibility in directing this development of autonomous machines—already research and production in this context creates a specific kind of future, and the people doing this should be aware of their responsibility. This is even more the case if they intentionally transfer responsibility for certain kinds of decisions onto machines. For some of these problems, more or less pragmatic legal solutions can be found. Not just these solutions—for example the introduction of a new kind of legal person—but also the development of designing machines for taking over specific decisions will have repercussions onto social and normative concepts such as responsibility.

Thus, discussing responsibility and potential legal status in the context of robotics means more than asking the question "Who is liable if something goes wrong?" and to introduce legal persons to solve problems based on the logic of the legal system only despite the effects this could have on society. It means to understand what happens if we intentionally hand over decision making onto machines. It means to leave room for decisions against machines taking over responsibility in specific contexts. It means to legally react to changing fundamental concepts and consciously create the space for these changes and to strengthen the awareness of the relevant institutions who will decide about the development of robotics.

References

[Ark08] Arkin. R. C. (2008). Governing lethal behavior: Embedding ethics in a hybrid deliberative/reactive robot architecture, GIT-GVU-07-11. https://smartech.gatech.edu/jspui/bitstream/1853/22715/1/formalizationv35.pdf.

[Bot14] Both, G., & Weber, J. (2014). Hands-free driving? Automatisiertes Fahren und Mensch-Maschine Interaktion. In E. Hilgendorf (Ed.), *Robotik im Kontext von Moral und Recht* (pp. 171–188). Baden-Baden: Nomos.

[Bos11] Boscarato, C. (2011). Who is responsible for a robot's actions? In B. van der Berg & L. Klaming (Eds.), *Technologies on the stand: legal and ethical questions in neuroscience and robotics* (pp. 383–402). Nijmegen: Wolf.

[Bro08] Brownsword, R. (2008). *Rights, regulation and technological revolution.* Oxford: Oxford University Press.

[EuR13] euRobotics. (2013). Suggestion for a green paper on legal issues in robotics. In Leroux, C., & Labrut, R. (Eds.). http://www.eu-robotics.net/cms/upload/PDF/euRobotics_Deliverable_D.3.2.1_Annex_Suggestion_GreenPaper_ELS_IssuesInRobotics.pdf.

[Fit13] Fitzi, G. (2013). Roboter als 'legale Personen' mit begrenzter Haftung. Eine soziologische Sicht. In E. Hilgendorf & J.-P. Günther (Eds.), *Robotik und Recht I* (pp. 377–398). Baden-Baden: Nomos.

[Hey13] Heyns, C. (2013). Report of the special rapporteur on extrajudicial, summary or arbitrary executions. http://www.ohchr.org/Documents/HRBodies/HRCouncil/RegularSession/Session23/A-HRC-23-47_en.pdf.

[Jon82] Jonas, H. (1982). Technology as a subject for ethics. *Social Research, 49,* 89182898.

[Wen09] Weng, Y. H., Chen, C. H., & Sun, C. T. (2009). Toward the human robot co-existence society: On safety intelligence for next generation robots. *International Journal of Social Robotics, 1,* 267–282.

Judgement

Amtsgericht München, Urteil vom 19.7.2007 – Az.: 275 C 15658/07, NJW RR 2008, 40.

Prospects of a Digital Society

How Artificial Intelligence Changes the World

Klaus Henning

Abstract The age of hybrid intelligence begins. The historical comparison between Gutenberg's book printing and the Artificial Intelligence revolution shows that fundamental disruptive innovations shake society to its foundations. We are in the middle of the shift that new dimensions of connectivity determine our lives and all the technical objects of the real world become intelligent. The success factors of human action for a sustainable digital transformation must therefore be agility, trust, and emotional awareness.

1 Introduction

Today's changes have a new driver that only played a secondary role: Artificial Intelligence. The question how this will influence our future world can be summarized in one sentence: *Machines become self-aware.*

This technological revolution will change and influence our entire world. I would like to summarize this in five theses:

1. The age of the digital universe is just starting: Everyone and everything will be intelligently connected.
2. There are no limitations: AI systems will enter *all* areas of this world. The age of machines with self-awareness begins.
3. Under these external conditions, a completely new type of human being is appearing. This includes a new understanding of relations between humans and machines. The age of the global-regional Homo Zappiens begins.
4. The new interactions of digital shadows and intelligent agents with humans and machines is the beginning of a new age of hybrid intelligence.
5. In order to manage and design this dimension of a sustainable digital transformation, design principles suitable for turbulences are required: Agility, trust, and (emotional) awareness.

K. Henning (✉)
IMA/ZLW & IfU, P3 OSTO, RWTH Aachen University, Aachen, Germany
e-mail: klaus.henning@ima-zlw-ifu.rwth-aachen.de

© Springer Nature Switzerland AG 2018
A. Karafillidis and R. Weidner (eds.), *Developing Support Technologies*, Biosystems & Biorobotics 23, https://doi.org/10.1007/978-3-030-01836-8_27

Fig. 1 Gutenberg's disruptive innovation: In ten years from the relic mirror to book printing (illustration based on [Mai16])

2 A Historical Comparison

In 1450, Johannes Gutenberg started dealing with book printing. It only took him ten years to get from a so-called relic mirror to mass book printing (Fig. 1) and another ten years later, there were printing houses all over Europe [Mai16]. In the year 1450, neither paper print nor rotation printing machines existed. Nobody knew the moveable letter yet. Only 10 years of development must have been a real innovation shock at that time.

Which were the success factors of Gutenberg? He was a man who lived against the Zeitgeist. He was very stubborn and invested a large amount of capital to reach his goal. He risked a lot: For example, he mortgaged his life insurance. The idea was more important than his own profit. And finally, Gutenberg was in love with scaling. He was a real production engineer. He knew how to use the basic innovation "reproducible letter" for his own purpose, the mass book printing. There is a certain tragedy in the whole story—and this should be a warning for us: His hometown Mainz was destroyed twenty years later during a war propelled for the first time in history by disseminating a massive number of printed leaflets.

3 The Age of the Digital Universe

Today we are facing a similarly fundamental disruptive innovation: Everyone and everything will be connected. The development into this direction has already begun a while ago, but with the entry of intelligence into these connected systems that is independent from the human mind, we are reaching a new dimension. The objects of our daily lives as well as vehicles and buildings become self-aware based on huge so-called "Big Data Lakes". We are surrounded by digital agents, digital twins, and digital shadows. It starts with smart phones that will become intelligent "personal assistants".

In addition, lots of agents closely connected to our personal lives will appear around us. We are about to have an interactive media center in our homes that takes care of the fridge, controls the entire energy supply, and reminds us of our breakfast. The "thinking" bumper of a car is still far away but it is foreseeable that we will be able to install intelligence into polymer materials so that a bumper will realize when a pedestrian is close and turn softer. Moreover, implanted cardiovascular pumps are going to become reality soon. Also, computers integrated into clothing will become normal. Perhaps we might even have a legal regulation for children to wear intelligent clothing on their way to school. There could be co-operative buildings in which humans intensively communicate with each other as well: Virtually open structures between apartments. At the moment, there is a research project of the Technical University Chemnitz with 200 elderly people in a huge tower block in Leipzig. One of the targets is to develop a kind of driver's license 4.0 that enables us to appropriately interact with the environment 4.0 [see http://nebeneinander-miteinander.de/ (2017)].

4   The Age of Machines with Self-awareness

Artificial Intelligence has in principle no limitations and will conquer all areas of this world. The age of machines that are self-aware starts now. But first of all: What is the core of Artificial Intelligence? What leads to its all-embracing influence? And why is "Deep Learning"—which is nothing more than closed-loop neuronal nets—the crucial breakthrough instrument for Artificial Intelligence, even though the theory about it has been invented already thirty years ago?

The decisive reason is the availability of data lakes—thanks to vast networks and digital infrastructures that are required for closed-loop neuronal nets to be developed effectively. The combination of a high-speed algorithm dealing with an enormous amount of data, a quite simple learning algorithm, and only a few amounts of "a priori" knowledge makes up the core of efficiency of modern Artificial Intelligence.

Thus, I would like to introduce the example of the "intelligent shoe" that already gets an identity in the moment it is ordered (Fig. 2, left). It knows what it is and it also knows its client. It knows what the client wants, for example whether the parameters of the client have to be monitored. And it knows what will be its condition and its route as well: It will have to make its way through the production plant in which a classic central control will not exist anymore. The production and transport units are in symbiosis with their own intelligent agents that negotiate with the intelligent shoe. This could happen "democratically" according to the political principle of separation and cooperation of powers—a method for which a first application for textile warp knitting machines already exists [Abb17].

Once the intelligent shoe is produced in that way, an automated transport unit will take further steps. Fully automated trucks are nothing new—a consortium of the

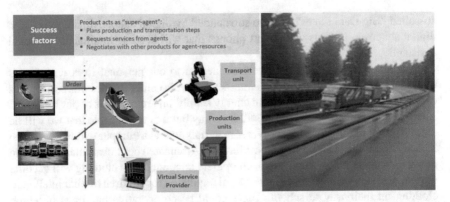

Fig. 2 Intelligent products acting as super-agents: How the intelligent shoe finds its way into the world and KONVOI driving on a German motorway. Adapted from an illustration of IMA/ZLW & IfU, RWTH Aachen University

University of Aachen (RWTH) and industry partners carried out the project KONVOI in 2009, in which fully automated trucks were tested on German highways for more than 5000 km in between the normal traffic flow (Fig. 2, right).

To sum up: Products will act as super-agents in the future. They plan their own production and transport. They establish requirements towards other agents, for example towards the production plant, and they negotiate with other agents about resources—on the streets or in production.

5   The Age of the Global-Regional Homo Zappiens

Over the last decades, all these developments have led to a new type of human that is called Homo Zappiens by the Dutch scientist Wim Veen Vee06] (Table 1). A completely new understanding of the relation between humans and machines is being generated. Wim Veen proposed this idea for the first time in 2006—at that time it still seemed quite visionary.

It is already normal that we have new forms of networking and that daily routines are organized over huge distances around the globe. Moreover, it became normal that we interact in a remote and virtual living and working environment. But what is new about the Homo Zappiens? He is able to multitask right from the start. He can think non-linearly. More dramatically: Today, a six-year-old can do 20 things simultaneously, but not one thing for a duration of five minutes. This is not bad, but rather good as this generation is already adapted to the conditions of information overload. It already possesses selection criteria and is processing information in a highly parallel fashion.

Table 1 Characteristics of the homo zappiens and homo sapiens in comparison

Homo Zappiens	Homo Sapiens
High speed	Conventional speed
Multi-tasking	Mono-tasking
Nonlinear approaches	Linear approaches
Iconic skills first	Reading skills first
Connected	Stand alone
Collaborative	Competitive
Learning by searching	Learning by absorbing
Learning by playing	Separating learning and playing
Learning by externalizing	Learning by internalizing
Using fantasy	Focused on reality

Adapted from [Vee06]

The Education Process Has to be Changed

This is what managers are radically confronted with. There will be a different generation of humans in our companies. This generation is different and wants to work differently because old structures seem obsolete to them. Learning by experience, events, and fun must be the trend if the universities and schools do not want to come to a dead end. Learning should always be fun. Children and adolescents intensively use social media of various kinds—such as freekickerz, Gronkh, BibisBeautyPlace, LeFloid, Emrah, Mr Wissen2go or especially TheSimpleClub. From a student's perspective, the last one is better than any school book.

The Change Will be Much More Radical Within the World of Business

Human work will be replaced or modified by the systems of Artificial Intelligence. It concerns white-collar jobs as well as highly-qualified work. IBM Watson AI computers can already take over some areas of controlling processes. Decentralized platforms will appear and even take over administrative tasks. This development will lead to an enormous economy of scales—not to mention autonomous systems in the air and on the street.

However, the fully automated car will not "only" drive fully automatically. It will also be, for example, the central digital twin of mobile nursing staff. Using the swarm intelligence together with its "colleague cars", it will be responsible for the entire disposition, documentation, traffic jam surveillance, route optimization etc. Thus, nursing staff can simply get into the vehicle and already start a conversation with the next patient via skype. In this case, the car becomes part of our everyday life as a social robot.

Fig. 3 A new dimension of communication between humans and intelligent objects. *Picture sources* IMA/ZLW & IfU, RWTH Aachen University, jim@Fotolia.com, P3 OSTO

6 The Age of Hybrid Intelligence

All in all, we can summarize that the human-to-machine interaction 4.0 creates a completely new dimension of cooperation between humans and intelligent objects. The age of hybrid intelligence between humans, machines, and their particular digital shadows and intelligent agents has begun (Fig. 3).

The age of predominance of the human beings over the objects created by them is coming to an end. Of course, the human-to-human interaction will still play a fundamental role in the future—perhaps even more than today. There is no alternative. For example, we increasingly get together far too late when it comes to talk about important things. One of the biggest absurdities of the last twenty years is the belief that human-to-human interaction can be done by e-mail. Some e-mail traffic should rather be forbidden and instead people should be forced to meet face-to-face—or at least via Skype—to settle disputes.

Furthermore, the human-to-machine interaction will keep on existing on different levels: on a screen, with a digital agent of the machine, or in immediate contact with the machine. But it is necessary to develop a "partnership on an equal footing". However, the idea of the "predominance" of the human over the machine is surely obsolete.

The machine-to-machine communication that takes places without any human will increase rapidly because every machine has its digital agent. The above-mentioned example of the textile warp knitting machine contains 200 software agents, but no SPS control anymore [Abb17]. Those agents even have a "right to vote" and choose "their" coordinators or speakers. Thus, digital shadows evolve, and people can interact with them, but within themselves they have a kind of "underground economy" of machine-to-machine communication.

In the Long Run, All Technical Objects of the Real World Will Become Intelligent

The digital shadow, the "digital skin" will become a dominant part of technology and human identity—in communication between machine and machine, between human and human and between machine and human. A "dynamics of digital shadows" will evolve that works simultaneously and develops progressively higher forms of intelligence.

This is the actual "revelation": Over the long term, all technical objects of the real world will become intelligent and develop self-awareness as well as a sense of themselves. There will be a life-long learning process of those objects and they will learn together with their technical partners and with the humans, too. "Driving schools" for technical objects will become normal. The omnipresent and discreet interaction between the digital shadows of technology and humans will dominate all aspects of communication.

This has enormous effects on the digital system landscape considering the dimensions that intelligence is everywhere, that the physical and digital world are linked and that we need new types of (IT)-infrastructures.

7 Success Factors of Human Action: Agility, Trust, and (Emotional) Awareness

What are the success factors for such a profound digital transformation that takes place under the dominant factor of Artificial Intelligence? According to our consulting experiences with P3 OSTO, the central success factors under these turbulent conditions are agility, trust, and (emotional) awareness (Fig. 4). In our experience, each of these factors is essential for the success of this transformation—they are "all-or-nothing factors". If one of the factors does not work, the whole thing will fail.

Agility does not only mean using a software development method like Scrum, but rather to adapt the entire structure of a company to agile principles—including all processes from product development, production, and product modifications to ramp-up processes and administrative structures. This is a huge issue and usually the product development department is ahead of the rest of the company. Central departments, such as administration, finances, or controlling often find it harder to implement agility even though it is more necessary in their departments than elsewhere.

Furthermore, agility requires a culture of trust—vertically and horizontally between humans and departments. If your organization does not provide this, you can forget about the transformation. It will not work.

And if there is no (emotional) awareness—if you do not ask: What is actually happening? What is culturally happening? Which tensions exist? If you do not have a perception for those factors in their range and diversity, you will not succeed. This way of being aware is the art to perceive but not to suppress the whole complexity and

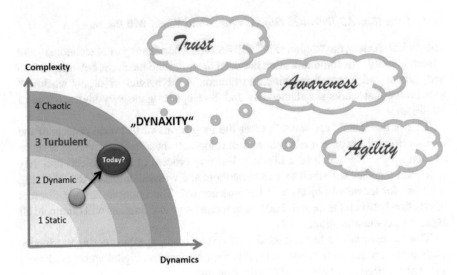

Fig. 4 Success factors of human acting under turbulent conditions. *Picture sources* P3 OSTO, Köln

dynamics (Dynaxity). But awareness is only an art to be learned. But mindfulness is only an art if you have learned to bear these perceptions and not rush into the reflex to rush to do something.

Each of these three features has to be fulfilled in high quality to make the change work, see [Hen15, Hen14]. May we all succeed in actively designing the new world of hybrid intelligence of humans and Artificial Intelligence according to our values, before others do it without reflecting their values.

References

[Abb17]   Abbas, B. (2017). *Verteilte Multi-Agenten zur Planung und Steuerung von Produk-tionsumgebungen auf Basis der Gewaltenteilung und Gewaltenverschränkung.* Aachen: Springer.

[Hen14]   Henning, K. (2014). *Die Kunst der kleinen Lösung - Wie Menschen und Unternehmen die Komplexität meistern.* Hamburg: Murmann Publishers.

[Hen15]   Henning, R. (2015). *Die Ego-Falle: 7 Möglichkeiten, Ihr Geschäft zu ruinieren.* Hamburg: Murmann Publishers.

[Mai16]   Mai, K.-R. (2016). *Gutenberg - der Mann, der die Welt veränderte.* Berlin: Propyläen Verlag.

[Vee06]   W. Veen, W., & Vrakking, B. (2006). Homo zappiens: Growing up in a digital age. London: Bloomsbury Publishing Plc.

Support in Times of Digitization

Athanasios Karafillidis and Robert Weidner

Abstract This paper sketches the connection of digital transformation and support technologies. Describing and conceptualizing human-machine configurations as "support" is intimately linked to material digitization processes. It is argued that support relations have to be considered as sensor-actuator-networks that consist of both organic and designed components embedded in a wider sociomaterial infrastructure. Interfacing various bodies then turns out to be the pivotal issue for support in the digital age. In this vein, challenges with regard to robots, automation, power, and collaboration are discussed.

1 Introduction

By the time when decision support systems came up in the late 1970s [Kee78], the notion of *support* has been more intriguing than the technical systems themselves. "Support" followed the ideas of "substitution" and "augmentation" as particular forms that describe our relations to technical artefacts, especially machines and automata [Car14, Mar15, Vis03]. Substitution obscures issues of responsibility; and augmentation is too focused on human capabilities and their shortcomings. In contrast, support now defines a different form of relationship between humans and technology with more latitude to negotiate and set the boundaries and roles between them.

Though simple tools may also be understood to support their human users, the explicitly communicated purpose to build support technology and the subsequent design and construction of relevant technology remains linked to the microprocessing digital computer. It is not accidental that the idea of technical support was put into words when digital computers could be programmed to be experts in a field

A. Karafillidis (✉) · R. Weidner
Laboratory of Manufacturing Technology, Helmut Schmidt University/University of the Federal Armed Forces Hamburg, Hamburg, Germany
e-mail: Karafillidis@hsu-hh.de

R. Weidner
Chair of Production Technology, University of Innsbruck, Innsbruck, Austria

© Springer Nature Switzerland AG 2018
A. Karafillidis and R. Weidner (eds.), *Developing Support Technologies*, Biosystems & Biorobotics 23, https://doi.org/10.1007/978-3-030-01836-8_28

with the purpose to support decisions of their human counterparts. The growth of interest for technical support and assistance in the past twenty years is a strong indicator that digitization does propel the preoccupation with support systems. Sketching various connections of support and digitization and naming critical challenges for future developments are the subject of this concluding chapter. It contends that sensor-actuator-networks consisting of both organic and designed components are the measure for most challenges of support relations.

2 The Matter of Digital Infrastructure

"Digitization" is a colorful but vacuous buzz word. Myriad phenomena are put under its auspices. Additionally, its narrative frame is mostly deterministic: digital technology has appeared due to general progress and now reshapes society. This is only half of the story, at most. Technology does neither appear out of the blue nor is it isolated when in operation [Mac99].

However, it is possible to flesh out some crucial material aspects of digitization without readily adopting the deterministic storyline attached to it. Paying attention to matter when describing digital society is not conventional at all. Software applications, big data, and powerful algorithms—or, as it were, in a generic sense: *Artificial Intelligence*—are seen to be the main drivers and get much more public attention these days. Yet their role is overemphasized. The reasons for this negligence of digital matter in debates about the digital age are not obvious. In the short run, exploiting the economic, political, scientific, or artistic possibilities of digital infrastructures appears to be more profitable at least. But digital society will not evolve just by devising even more powerful deep learning algorithms, optimizing neural networks, and gathering even more data than before. Digitization will have to incorporate techno-material innovations to advance substantially.

Shifting the attention to the techno-material infrastructure proper, digitization reveals itself as a close combination of three intertwined developments since the 1940s: the digital computer and its architecture, the exponential growth and distribution of computing speed enabled by the microprocessor, and the global connectivity achieved by setting up computational networks [Pra15]. Digital computers, microprocessors, and connectivity are crucial techno-material prerequisites for what is called digitization. But the watershed is marked by a different aspect.

Digitization as we know it did not emerge until all kinds of sensors and actuators were embedded into this infrastructure. Algorithms were there before, but their power became discernible and prevalent, when they could also select and modify sensors, actuators, and their connections. There was data abound since the advent of digital computers, microprocessors, and the internet, but data became really big when mobile sensors started to pervade society. This sensor-actuator-embedding into the existing digital infrastructures is responsible for the hopes as well as the fears, the business opportunities as well as the surveillance risks, and the desired comfort as well as the authoritative control that are being discussed. The age of context sensitivity [Sco14], the granularity of digitization [Kuc14], and the relevant critique [e.g., Kee18] are a result of this techno-material insertion of sensors and actuators into networks on a huge scale and a fast pace.

3 Support: Hybrid Sensor-Actuator Networks

The idea to develop support technology is historically an immediate offspring of this technological embodiment of digitization. More precisely, conceiving the interaction between humans and artefacts as support relation became an option when technical sensors and actuators were aligned with human sensory and motor abilities in a digital infrastructure. Technical sensors and actuators respond to human behavior or parts of human bodies. Humans, in turn, ascribe behavior and responsiveness to digital machines, that/who started to support decisions, work and production, the organization of our daily lives, and also bodily and organic functions by now. In effect, humans and technology are actually getting closer.

Since support relations have become a real option, the gap between completely manual tasks and activities on the one hand and fully automated ones on the other is vanishing. A hybridization of biomechanical and technical elements has become common: human sensorimotor capacities are interfaced with technical sensors, actuators, and kinematics. "Sensors" and "actuators" do not only indicate components of technical artefacts in this description (maybe "actuators" should better be called "effectors" which is a composite of actuators, kinematics, and drivers, yet we keep the term—being well aware that it has a more restricted meaning in engineering). Technical sensors and actuators are crisscrossed with elements of human sensorimotor systems to generate integrated hybrid systems (Fig. 1). This is one of the main reasons why support systems cannot be reduced to an engineering issue. Another one is the embeddedness of these relations of Fig. 1 into further networks and institutions—the "seamless web" of societal technology construction [Bij87].

Take an example. A passive exoskeleton may be built without any technical sensors. But it is wrong to conclude that the support system lacks sensors in this case. The exoskeleton works in connection with the human senses/sensors of its wearer and the designed loose or strict coupling of its actuators and kinematics to the motor activity of the human body. The network of components that constitutes support transcends the "passive" artefact. The technical system is moreover contingent on a due

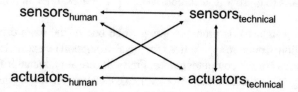

Fig. 1 Possible relations in a minimal hybrid sensor-actuator network for support. Not all relations need to be realized but at least one of the indicated relations in the middle is mandatory

coupling to the perceptions and actions of the wearer's peer groups. When the hybrid sensor-actuator network fails, then the exoskeleton fails, too—even if it is highly praised, mature, and elaborate from a purely technical point of view. This could also be shown for (lightweight) robots, AR and VR systems, social robots, health trackers, smart watches, or simple lifting aids and tools. Indeed, classical fenced industrial robots can also be considered to provide support on different levels: e.g. for humans, production processes, whole enterprises, or the economy. Attending to hybrid sensor-actuator networks of support shifts perspective even in this case. The notion of cyber-physical systems is close but misses a crucial point, unless "cyber" and "physical" are *both* conceived as being *both* organic and technical.

However, the new proximity and hybridity of sociotechnical human-artefact assemblies becomes more salient and vivid when we stick to the situated and local support of human beings and take recent developments into account.

4 Major Challenge: Interfacing Bodies in Digitized Worlds

The example of the exoskeleton is not chosen haphazardly. In our opinion, a major challenge of developing support technologies lies in *physical* support systems, for example wearable technology. This reflects part of our own research interests but there is more substance to this conjecture than simply overrating the own research agenda. The relevant but hardly stated question is: *how will we take our bodies with us into the digital age?* After all, we will have to count and calculate on them for some time to come—longer than some technocentristic transhumanism may suppose.

The human body is a blind spot of digitization. Public discussions about digitization mostly ignore the body and prefer to highlight connectivity, algorithms, apps, data mining, or touchscreen functionality. "Cognitive" support is ubiquitous and obviously much easier to implement right now. Touchscreen devices are widespread and have found their way into our everyday lives. In contrast, robots and other physical support systems like all kinds of exoskeletons are just starting to leave the labs. Augmented and virtual realities are still looking for suitable ways to integrate the bodies of users. Certainly, there are many projects seizing on relevant solutions. But they still have to prove that they can live up to their promises—beyond some all too enthusiastic portrayals of designers, engineers, tech bloggers, and journalists. However, huge progress has been made in this respect, but it is definitely still much untapped potential regarding the integration of human bodies as well as artificial body parts (e.g., robots, wearable robots, prostheses, implants) into the digital infrastructure.

The gaming industry is probably going to be one of the main drivers for such body integration. Concrete applications for physical support by exoskeletons or smart textiles and suits are VR computer games. First suits are already mature enough to be marketed (e.g. https://teslasuit.io). However, exoskeletons are mostly developed (and distributed) for nursing, military, industrial, and rehabilitation purposes [Hoc15]. Their public appearance is often connected to demonstrations how they make people with paraplegia walk again—for rehabilitation training only, to be sure. The digital integration of the body does also occur with small devices and implants. Hearing systems do currently evolve to be more than aids for hearing-impaired people, but rather present a real possibility for a volitional design of the individual acoustic environment and for interfacing with other applications. Experiences with medical implants (e.g., pacemakers, cochlear) have a longer history but the possibilities of their digital connectivity are still to be explored.

Interfacing of human bodies with artificial bodies or body parts will remain an issue for a long time to come. Questions about comfort play a significant role for acceptance. Hence, the design and construction of various accurate interfaces is key. Hardwiring prostheses, smart clothing, or exoskeletons with the human body—either for immediate material connection or signal processing or both (e.g. adaptable molds and straps for extremities, electromyography for control via muscular activity, or inertial sensors for movement and stature recognition)—will be further refined and continue to generate new interfacing possibilities.

All this is accompanied and reinforced by the development of interfaces for mutual connectivity of technical objects. Technical support systems can be coupled to each other directly or via people using them in parallel. For example, smart tools recognize movements and positions in space to support certain tasks; or their users sense the need for more support and might be able to control, e.g., an exoskeleton with another tool that they are using. The Internet of Things (IoT), cyber-physical systems, or the German term "Industrie 4.0" are the generic indications for these fine-grained developments.

5 Challenging Support Issues: On Robots, Automation, Power, and Projects

Thinking about future possibilities of support in a digital society also includes accounting for developments that might turn out as detrimental to communal life but also to the capacity to turn engineered inventions into societal innovations. There is no intention to present a complete list, which is actually impossible, but to simply present a few points that are a little more concrete than fear mongering of machine control over people or the no doubt serious issues regarding privacy, data security, or the surveillance state. The main perspective is still physical or corporeal support.

5.1 Autonomous Robots

Support is certainly not confined to wearable solutions or ambient technologies like, e.g., smart homes. Autonomous robots are supporting people, too. Many of today's technical challenges are situated in the fascinating field of robotics. Yet there is also a bias with regard to what robots can currently do and in which contexts they can be employed in a meaningful way. For example, autonomous and basically humanoid, wheeled robots are considered as future assistants at home, in public spaces (e.g.

museums), and particularly in elderly care. This is the vision for the future of nursing propagated by the developing engineers of such robots and many tech-journalists. There is nothing wrong with having this vision. But since it is a vision, it should be treated like one. Else it could quickly become an obstacle for developing technical support in elderly care that is truly viable.

Nursing robots are an exemplary case for technology development that is mostly technology-driven—that is, without much interest in what people actually want and need. Anybody who has examined the work processes, skill, and dexterity of caregivers immediately realizes that current robots are not able to do this job in near future. Even single tasks cannot be supported efficiently. Autonomous nursing robots and their capabilities are therefore overrated. They are not what caregivers really want, they are neither what caretakers really need, and they are not even a sound economic solution for nursing institutions right now. In the current state of affairs, robots will not substitute nursing home professionals for many years to come—unless it is politically enforced without an interest in what the people involved demand.

In general, fearful reports about robots taking over elderly care must be distinguished from the actual performance abilities of robots, which are poor regarding the inevitable and necessary adaptability to shifting and dynamic environments, the multi-dimensionality of many tasks, and the required pace of learning. These are severe operational issues. They differ from the also valid complaints of some experts about robots lacking compassion and the experience of pain, which makes them ineligible for tasks that involve autonomous work on sentient beings. The technical and political vision of solving problems of demographic change by introducing robots into caregiving therefore distracts attention and financial resources from acceptable (technical as well as non-technical) solutions.

These reservations have to be taken as specific as they are. They do not reject robots as a species but the inappropriateness of certain visions and imaginaries that could easily create irreversible, expensive, and detrimental path dependencies. In contrast to caring robots, the possibilities of social robots built for play, entertainment, and pastime are rather underrated in the field of care. For example, care professionals argue that such robots are only pretending to have emotions and that this is an unethical deception of older people. But if social robots trigger positive emotions in older people and add moments to their lives in which they forget their loneliness and boredom [Bed17], then the argument opposing this kind of support as not being "real" is untenable and to some extent even irresponsible. In this vein, social robots can actually support (definitely not: substitute) the overworked caregivers.

5.2 Automation and Substitution

This brings us to a central issue of many debates concerning the digital age: the fragile distinction between support and substitution. Providing "support" instead of automating a process and thus substituting the role of human individuals represent two different approaches to technology development. Substitution is about automation. In contrast, support is about an integration of heterogeneous elements. It tries to reinstall, preserve, protect, prevent, or augment human activities by incorporating technical systems. It is oriented towards demands and needs of humans in particular, albeit support might also refer to organizational demands. However, there may be concerns that support technologies could just be a precursor for a subtle transfer of decision authority and activities to machines and the ones who control them. This could ultimately lead to an incremental substitution of humans or to a more thorough exploitation of staff to achieve higher productivity. This is no doubt a possible scenario, in particular within for-profit organizations. Yet supplanting the generic idea of Human-Machine Interaction with the notion of support relations on a conceptual level has at least two distinctive effects: it helps to create technology that is sensitive to sociomaterial human particularities; and it will necessitate to discuss in each case, whether some invention is going to support or substitute human activities and whether a substitution is intended or not. The concept of support thus provides a common ground for adequate collective judgments regarding the effects of automation.

Elon Musk, well known tech-entrepreneur and CEO of Tesla, tweeted a response on April 13th, 2018, that the production problems they encounter with Model 3 are due to excessive automation. This point addresses exactly the issue to which the support paradigm is a response: there is obviously a need to reconsider and actively decide under which circumstances automation und thus substitution of the human workforce does make sense—even in industries like automobile production where automation is self-evident and unquestionably considered to be beneficial, more efficient, and more effective. Musk also adds that "humans are underrated". Obviously, Tesla would have profited from a discussion alongside the distinction of support and substitution. Once the automation decision is made and implemented, it becomes an uphill battle to strip it down again and bring people back in.

The relation of support and substitution is non-linear and multilayered though. There is no one-best-way and no decision that is correct once and for all. First of all, it depends on what and whom is to be supported. Furthermore, support can be achieved by substitution. Just to give one example: Websites already support customers to customize the ordered products while the corporation substitutes its workforce for automated processes that run these websites triggering production machines and automated dispatch. In general, who is supported by whom and where and how substitution may take place when some technology prevails is an open question. Who is supported when a data pill is employed? The one who swallows it, the physician in charge, or the insurance company? Maybe all of them but maybe none. Does it supplant a function, an activity, or a task? To find any answers, the conditions of any singular case have to be examined closely, both for sound scientific analysis and practical decision making.

5.3 Power: Energy and Inequality

A complex of further challenges can be subsumed under the label of power. This includes both physical energy (production, storage, and consumption) and socio-economic power (capital, status, and inequality). These two can be considered separately of course, but there is one point of articulation here: energy is not a collective good and digitization increases the individual energy demand. Global power elites are able to spend more to get the energy they need to harness the potential of digital infrastructure and to realize their goals. Additionally, they do also have more potential to produce energy themselves by investing capital to buy, for example, solar panels, which again require property to install them. Inequality is rising in the digital society and the energy issue will be one of its facilitators.

Energy is also a great engineering challenge. Its efficiency is of global importance. The worldwide employment of technologies heavily depends on the production of huge amounts of energy that in turn depends on deploying technologies that again consume huge amounts of energy and additionally produce externalities. This dependence of technology on further technology has become absolute [Luh00, pp. 378–379].

Energy consumption and battery life is and will obviously remain an issue for the development of any active support system. Anyway, battery research to optimize output and storage capacity must be complemented with further inquiry into alternative forms and solutions for energy provision (e.g., pressurized gases, shape memory alloy, the efficiency of solar panels, etc.). But any progress in this direction would be worthless when the appropriate drivers are not developed in accord.

Of course, the problem of inequality is not only coupled to energy consumption and production. It also concerns the purchase of support technology, that is, its unequal distribution. Technical support will become a question of income (as is health care) [Alt15]. Getting the gear suitable for the own demands, business interests, or ailments will also be a question of access, which includes spending capacity but is not confined to it. Access is not settled with money only. It also heavily relies on the conditions of regional digital infrastructures. This will necessarily fuel a further concentration of people in urban areas and thus reinforce existing regional inequalities and a further concentration of population and activities in metropolitan areas [Cas10].

In principle, support technology has the potential to compensate defects and to foster equality. Spectacles for eye support, to give a really simple and non-digital example, prevent severe social divisions between the "focused" and the "blurred" people. This implies that technical solutions enable participation. Yet their potential for separation is equally high. As with any invented and disseminated technology it comes to political decisions to stipulate how accessibility and deployment are organized and controlled. Additionally, it points to the issue of defining societal standards. What would be the standard for compensation with respect to human senses and bodies? There is on definite answer. Compensating differences is not desirable in itself. Social control adds to these questions about the power to define standards. The connectivity of current technical devices allows for much debated governmental control possibilities but also for different forms of control in peer groups and families. Yet control is no a zero-sum game. Individual control opportunities rise in the same amount. The control society described by the philosopher Gilles Deleuze [Del90] is not so much about unilateral domination but about the continuation of mutual efforts at control.

5.4   Collaboration in Projects

Discussing these societal issues is of great importance for developing support tech-nologies, even if it seems far removed from the everyday processes of technical development at first. But ignorance and reservation with respect to earth and social inequalities are not an option [Lat17]. Engineers and all other developers involved in such processes, must be aware of these connections—not only for ethical, "social", and legal reasons but rather additionally for the very success of the project and the invented technology itself. Confining acceptance issues only to individual psycholog-ical factor models belongs to the past. The challenge is to build support technologies that are affordable and attentive to the political issues of control and inequality.

How can this be achieved? Once again, there are no general success factors. Yet there seems to be no alternative to interdisciplinary collaboration [Led15]. In this kind of projects, expertise must be networked and distributed. Expertise is not necessarily brought to the project but rather its main outcome. In addition, collaboration is extended to comprise people who are supposed to use and profit from the developed technology. Since this transcends the interdisciplinary academic team and introduces citizens or corporate natives into the process of development, such an approach is understood to be transdisciplinary. This is not exactly "citizen science" but certainly belongs to a similar domain.

The future challenge lies in testing and evaluating possible forms of interdisci-plinary work and participation. Any project has its peculiarities with respect to the disciplinary mix in the team and the potential users and beneficiaries of the technical devices to be designed and constructed. Finding a form of collaboration to develop technology that people really want is absolutely fundamental. It takes its time and needs some latitude—and this is also the biggest obstacle for establishing such teams and for daring to proceed in this direction. Due to much more iterations, continuous testing in the field, and the required exchange and negotiations among the multiple perspectives of the involved parties the development process is slower in transdisci-plinary projects [Bro15]. For Research & Development departments this may appear too cumbersome. Stakeholders and investors need to comprehend that such projects cannot be measured by delivering a predefined solution on time. No doubt a result is mandatory. But the assessment must also put a premium on the process itself and the accumulated competence.

Another aspect of collaboration is the challenge of participation. Like before, it can be boiled down to a time issue. Finding the right people and partners for development is demanding, notably it takes a lot of time to coordinate meetings and activities. But time is also the main factor in a further respect. Some support technologies are developed for prevention of illness and disease. Yet prevention is hardly an incentive for acceptance in participatory development. Prevention refers to an uncertain event in some distant future that seems irrelevant from today's perspective. Thus, other incentives are needed that might nudge people in the relevant preventive practices. Gamification is an option. Convenience is another. Still there is much ignorance how to manage that. However, serious participation of users from the outset generates experts outside academia and also a certain excitement that motivates them to talk about it. Peers have a huge influence and the form of communication surrounding and permeating the technology is key to participatory development and acceptance.

6 Conclusion

This concluding chapter has argued that there is an immediate link between the idea of support technology and digitization. It has presented a simple model of hybrid sensor-actuator networks that can be used to expound this connection between digital transformation and support. From there certain fields have been sketched that could turn out to be critical for the future development of support systems. This leads us back to the books original purpose: showing that support has the potential to re-design the classic notion of Human-Machine Interaction. We contend that this potential is harnessed only when we succeed to integrate multiple perspectives and multiple bodies to design, construct, deploy, and evaluate support systems. If the objective is to create technology that people really want, this kind of integration could be the best bet right now.

References

[Alt15]   Altman, R. (2015). Distribute AI benefits fairly. *Nature, 521,* 417–418.
[Bed17]   Bedaf, S, Huijnen, C., van den Heuvel, R., & de Witte, L. (2017). Robots supporting care for elderly people. In P. Encarnação & A. M. Cook (Eds.), *Robotic assistive technologies. Principles and practice* (pp. 309–332). Boca Raton: CRC Press.
[Bij87]   Bijker, W. E., Hughes, T. P., & Pinch, T. J. (1987). *The social construction of technological systems. New directions in the sociology and history of technology.* Cambridge: MIT Press.
[Bro15]   Brown, R. R., Deletic, A., & Wong, T. H. F. (2015). How to catalyse collaboration. *Nature, 525,* 315–317.
[Car14]   Carr, N. (2014). *The glass cage. How our computers are changing us.* New York: W. W. Norton.
[Cas10]   Castells, M. (2010). Globalisation, networking, urbanisation: Reflections on the spatial dynamics of the information age. *Urban Studies, 47*(13), 2737–2745.
[Del90]   Deleuze, G. (1990, May). Post-Scriptum sur les sociétés de contrôle. *L'autre journal, 1.*
[Hoc15]   Hochberg, C., Schwarz, O., & Schneider, U. (2015). Aspects of human engineering—Bio-optimized design of wearable machines. In A. Verl, A. Albu-Schäffer, O. Brock, & A. Raatz (Eds.), *Soft robotics. Transferring theory to application* (pp. 184–197). Berlin: Springer.
[Kee78]   Keen, P. G. W., & Scott-Morton, M. S. (1978). *Decision support systems: An organizational perspective.* Reading, MA: Addison-Wesley.
[Kee18]   Keen, A. (2018). *How to fix the future. staying human in the digital age.* London: Atlantic Books.
[Kuc14]   Kucklick, C. (2014). *Die granulare Gesellschaft.* Berlin: Ullstein.
[Lat17]   Latour, B. (2017). *Où atterrir? Comment s'orienter en politique.* Paris: La Découverte.
[Led15]   Ledford, H. (2015). Team Science. *Nature, 525,* 308–311.

[Luh00]   Luhmann, N. (2000). *Organisation und Entscheidung*. Wiesbaden: Westdeutscher Verlag.

[Mac99]   MacKenzie, D., & Wajcman, J. (Eds.). (1999). *The social shaping of technology* (2nd ed.). Buckingham: Open UP.

[Mar15]   Markoff, J. (2015). *Machines of loving grace. The quest for common ground between humans and robots*. New York: Harper Collins.

[Pra15]   Pratt, G. A. (2015). Is a Cambrian explosion coming for robotics? *Journal of Economic Perspectives, 29*(3), 51–60.

[Sco14]   Scoble, R., & Israel, S. (2014). *Age of context. Mobile, sensors, data and the future of privacy*. USA: Patrick Brewster Press.

[Vis03]   Viseu, A. (2003). Simulation and augmentation: Issues of wearable computers. *Ethics and Information Technology, 5,* 17–26.

Index

© Springer Nature Switzerland AG 2018
A. Karafillidis and R. Weidner (eds.), *Developing Support Technologies*, Biosystems &
Biorobotics 23, https://doi.org/10.1007/978-3-030-01836-8

Printed in the United States
By Bookmasters